T0134936

Studies in Computational Intelligence

Volume 945

The series "Studies in Computational Intelligence" (SCI) publishes new developments and advances in the various areas of computational intelligence—quickly and with a high quality. The intent is to cover the theory, applications, and design methods of computational intelligence, as embedded in the fields of engineering, computer science, physics and life sciences, as well as the methodologies behind them. The series contains monographs, lecture notes and edited volumes in computational intelligence spanning the areas of neural networks, connectionist systems, genetic algorithms, evolutionary computation, artificial intelligence, cellular automata, self-organizing systems, soft computing, fuzzy systems, and hybrid intelligent systems. Of particular value to both the contributors and the readership are the short publication timeframe and the world-wide distribution, which enable both wide and rapid dissemination of research output.

Indexed by SCOPUS, DBLP, WTI Frankfurt eG, zbMATH, SCImago.

All books published in the series are submitted for consideration in Web of Science.

More information about this series at http://www.springer.com/series/7092

Khaled R. Ahmed · Aboul Ella Hassanien
Editors

Deep Learning and Big Data for Intelligent Transportation

Enabling Technologies and Future Trends

Editors
Khaled R. Ahmed
School of Computing
Southern Illinois University
Carbondale, IL, USA

Aboul Ella Hassanien
Information Technology Department
Faculty of Computers and Artificial Intelligence
Cairo University
Giza, Egypt

Scientific Research Group in Egypt (SRGE)
Giza, Egypt

ISSN 1860-949X ISSN 1860-9503 (electronic)
Studies in Computational Intelligence
ISBN 978-3-030-65663-8 ISBN 978-3-030-65661-4 (eBook)
https://doi.org/10.1007/978-3-030-65661-4

This Springer imprint is published by the registered company Springer Nature Switzerland AG
The registered company address is: Gewerbestrasse 11, 6330 Cham, Switzerland

Preface

Deep learning and big data are very energetic and vital research topics of today's technologies. This book contributes to the progress toward intelligent transportation, such as fully autonomous vehicles. It emphasizes new data management and machine learning approaches, such as big data, deep learning, and reinforcement learning. Road sensors, UAVs, GPS, CCTV, and incident reports are sources of massive amounts of data crucial to making serious traffic decisions. Herewith this substantial volume and velocity of data, it is challenging to build reliable prediction models based on machine learning methods and traditional relational databases. Therefore, this book includes 12 chapters that elaborated the recent research works on big data, deep convolution networks, and IoT-based smart solutions to limit the vehicle's speed in a particular region, to predict steering angle, to control traffic signals, to support autonomous safe driving, to count crowds ,and to detect animals on roads for mitigating animal–vehicle accidents. This book would serve broad readers, including researchers, academicians, students, and working professionals in vehicle manufacturing, health and transportation departments, and networking companies. The structure of the book reflects an effort to present the individual chapters coherently and logically. The separate chapters are grouped into three main areas: big data and autonomous driving, deep learning and object detection for safe driving, and AI and IoT for intelligent transportation. An abstract and several keywords at the beginning of each chapter support a reading overview.

Big Data and Autonomous Vehicles

Chapter "Big Data Technologies with Computational Model Computing Using Hadoop with Scheduling Challenges" mainly focused on big data (BD) qualities and profoundly examined the complications raised by BD registering frameworks. The authors have clarified the estimation of big data mining in different areas. This chapter elaborates on the challenges incorporated with big data analytics (BDA) that are classified into data challenges, process challenges, and management challenges. Data challenges are a group of challenges related to data characteristics, such as volume, velocity, veracity, and visualization. Challenges that may be encountered while processing the data, such as data analysis, modeling, and data aggregation, are called process challenges. However, tackling privacy, security, operational cost, data governance, and data sharing are considered management challenges. This chapter gives a definite knowledge into the design, systems, and practices that are now continued in the big data process.

Chapter "Big Data for Autonomous Vehicles" demonstrated the significance, applications, related technologies, issues, and research directions in big data for autonomous vehicles (AVs). It presents the protocols and standards for vehicle-to-everything (V2X) communication. Moreover, this chapter discusses the on-going research concerning the technologies involved in big data processing in AV, such as data transmission, acquisition, storage, computing, analytics, and processing. Finally, research directions, open research issues in big data, machine learning for vehicular networks, and vehicular network security are presented.

Deep Learning and Object Detection for Safe Driving

Chapter "Analysis of Target Detection and Tracking for Intelligent Vision System" discussed the concept of an intelligent video surveillance system for smart transportation and the significant challenges to detect, classify, and track moving objects (e.g., vehicles, animal, people, etc.). The significant challenges that affect the performance to track/detect objects are elaborated, such as illumination changes, weather conditions, dynamic and clutter background, occlusion, moving cast shadow, and images noises due to camera jitter. This chapter presents the potential algorithms for efficient object detection and tracking, such as temporal difference method, simple background subtraction method, optical flow method, background model approaches, segmentation, and feature-based approaches.

Chapter "Enhanced End-to-End System for Autonomous Driving Using Deep Convolutional Networks" discussed the steering algorithms for autonomous driving based on deep convolution networks. The authors proposed independent steering algorithms to predict steering angle, analyze road markings, determine the road radius of the road with lane deviation, and identify objects. This chapter used Yolov-V3 based on darknet-53 architecture. The authors trained using COCO dataset, having 80 labels such as a person, car, and bus. The proposed algorithm able to suggest the driver either is it safe to accelerate the vehicle or not. It may also help to warn drivers if there is any deviation from the lane center. This algorithm could be used by police officers to detect vehicles with their velocities while driving.

Chapter "Deep Learning Technologies to Mitigate Deer–Vehicle Collisions" demonstrated the importance of the deer–vehicle collisions (DVCs) global problem and explored various traditional DVCs mitigation techniques. The authors developed a deep learning algorithm based on Yolov5 to predict deer on roads. The generated prediction model achieves about 99.5% mAP to detect deer in various weather conditions.

Chapter "Night-to-Day Road Scene Translation Using Generative Adversarial Network with Structural Similarity Loss for Night Driving Safety" proposed an enhanced deep learning algorithm to reduce the risk of traffic accidents at night by providing motorists a daytime view while driving at nighttime. The authors developed an algorithm based on generative adversarial networks (GAN) to translate the nighttime road scenes into daytime ones while preserving the other road conditions. To improve the perceptual image quality, the authors introduced additional structural similarity loss. The qualitative and quantitative measurements suggested the proposed algorithm's ability to transform the nighttime image into the daytime scene with much sound quality.

Chapter "Safer-Driving: Application of Deep Transfer Learning to Build Intelligent Transportation Systems" demonstrated comprehensive guidance on using a transfer learning strategy to generate a detection model to establish a safer transportation system. The author discussed a case study to detect deer on the road by implementing a deep learning algorithm based on a pre-trained CNN network (VGG16). The developed model performed about 0.916 accuracy on the testing dataset.

Chapter "Leveraging CNN Deep Learning Model for Smart Parking" leveraged the convolution neural network (CNN) deep learning model for smart parking. The authors trained the CNN algorithm to automate car parking to find empty and filled parking slots. The generated parking slot detection model's efficiency is derived from the outcome of binary classification and recalling of the precision metrics.

Chapter "Estimating Crowd Size for Public Place Surveillance Using Deep Learning" discussed using deep learning algorithms for automatic surveillance detection of any abnormal activities in the crowd. The authors explored four major convolution neural network (CNN) architectures used for crowd counting switch CNN, multicolumn CNN, CSRNet, and cascaded CNN. They performed performance analysis for these models over different types of datasets containing images of varying crowd densities (high, medium, and low). Finally, this chapter concludes that CSRNet is the most promising network to count crowd.

AI and IoT for Intelligent Transportation

Chapter "IoT Based Regional Speed Restriction Using Smart Sign Boards" mainly focused on improving safety on our roads by avoiding fatal accidents caused by speeding. The authors proposed an Internet of things (IoT)-based smart solution that automatically controls vehicles' speed and customizes the regional speed limit. They explored the current solutions and challenges that track and enforce the speed limit to prevent road accidents on our roads. Also, they implemented a system prototype to control the speed of the vehicle using Arduino microprocessor, Bluetooth low energy, and several sensors such as speed sensors, and cameras. Finally, the authors draw the conclusion, future work, and provided recommendations to automobile manufactures.

Chapter "Synergy of Internet of Things with Cloud, Artificial Intelligence and Blockchain for Empowering Autonomous Vehicles" elaborated on the uses of the Internet of things (IoT), artificial intelligent (AI), cloud, and blockchain technologies for developing autonomous vehicles. It provides a detailed discussion of the challenges of IoT in autonomous vehicles. This chapter explored and exposed the synergy of IoT with AI, cloud, and blockchain. The IoT and AI transform vehicles into fully smarter to detect the vehicle's environment and make appropriate decisions accordingly. However, IoT and cloud are used to store and process collected data and enable communication among vehicles and roadside units with minimal cost. To secure the interactions among vehicles communicated through integrated sensors, blockchain and IoT are used. Blockchain is used to track the autonomous vehicles' interactions to secure transactions such as ride-sharing transactions and toll transactions. This chapter discussed several IoT applications of AI, cloud, and blockchain in autonomous vehicles.

Chapter "Combining Artificial Intelligence with Robotic Process Automation—An Intelligent Automation Approach" provided insights about the new intelligent automation approach based on AI integration with RPA in a smart transportation system. This chapter discussed the process automation that brings significant benefits for businesses and organizations characterized by rich data flow and information-intensive. Robotic process automation (RPA) is used and integrated with cognitive techniques such as machine learning, speech recognition, and natural language processing to automate higher-order tasks. Thus, this chapter explored the technologies required for a successful combination of RPA and artificially intelligent.

Finally, we hope the book may inspire researchers interested in the challenges and complexities of rapidly changing big data and deep learning for intelligent transportation. The editors would like to express their gratitude to authors and reviewers for their hard works and efforts to succeed in this productive research project.

Carbondale, USA
Giza, Egypt

Khaled R. Ahmed
Aboul Ella Hassanien

Contents

Big Data and Autonomous Vehicles

Big Data Technologies with Computational Model Computing Using Hadoop with Scheduling Challenges 3
E. B. Priyanka, S. Thangavel, B. Meenakshipriya, D. Venkatesa Prabu, and N. S. Sivakumar

Big Data for Autonomous Vehicles 21
Rinki Sharma

Deep Learning and Object Detection for Safe Driving

Analysis of Target Detection and Tracking for Intelligent Vision System 51
K. Kalirajan, K. Balaji, D. Venugopal, and V. Seethalakshmi

Enhanced End-to-End System for Autonomous Driving Using Deep Convolutional Networks 81
Balaji Muthazhagan and Suriya Sundaramoorthy

Deep Learning Technologies to Mitigate Deer-Vehicle Collisions 103
Md. Jawad Siddique and Khaled R. Ahmed

Night-to-Day Road Scene Translation Using Generative Adversarial Network with Structural Similarity Loss for Night Driving Safety 119
Igi Ardiyanto, Indah Soesanti, and Dwiyan Cahya Qairawan

Safer-Driving: Application of Deep Transfer Learning to Build Intelligent Transportation Systems 135
Ramazan Ünlü

Leveraging CNN Deep Learning Model for Smart Parking 151
Guruvareddiyur Rangaraju Karpagam, Abishek Ganapathy, Aadhavan Chellamuthu Kavin Raj, Saravanan Manigandan, J. R. Neeraj Julian, and S. Raaja Vignesh

Estimating Crowd Size for Public Place Surveillance Using Deep Learning 175
Deevesh Chaudhary, Sunil Kumar, and Vijaypal Singh Dhaka

AI and IoT for Intelligent Transportation

IoT Based Regional Speed Restriction Using Smart Sign Boards 201
P. Madhumathy, H. K. Nitish Kumar, Pankhuri, and D. S. Supreeth Narayan

Synergy of Internet of Things with Cloud, Artificial Intelligence and Blockchain for Empowering Autonomous Vehicles 225
C. Muralidharan, Y. Mohamed Sirajudeen, and R. Anitha

Combining Artificial Intelligence with Robotic Process Automation—An Intelligent Automation Approach 245
Nishant Jha, Deepak Prashar, and Amandeep Nagpal

Big Data and Autonomous Vehicles

Big Data Technologies with Computational Model Computing Using Hadoop with Scheduling Challenges

E. B. Priyanka, S. Thangavel, B. Meenakshipriya, D. Venkatesa Prabu, and N. S. Sivakumar

Abstract Big Data (BD), with their capability to learn esteemed bits of knowledge for an improved dynamic cycle, have as of late pulled in generous enthusiasm from the two scholastics and specialists. Big Data Analytics (BDA) is progressively turning into a moving practice that numerous associations embrace to build significant data from BD. The examination cycle, including the sending and utilization of BDA instruments, is seen by associations as a device to improve operational effectiveness; however, it has vital potential, drives new income streams, and increase upper hands over business rivals. Be that as it may, there are various sorts of expository applications to consider. In this manner, preceding rushed use and purchasing expensive BD instruments, there is a requirement for associations first to comprehend the BDA scene. Given the BD and BDA's fantastic idea, this paper presents a state-of-craftsmanship survey that gives an all-encompassing perspective on the BD difficulties, and BDA techniques speculated/proposed/ utilized associations to help other people comprehend this scene to settle on strong venture choices. The examination introduced in this part has recognized significant BD research considers contributing both adroitly and precisely to the extension and gathering of scholarly riches to the BDA in innovation and hierarchical asset the board discipline. While there are a few productive methodologies for exhibiting MapReduce outstanding

E. B. Priyanka (✉)
Department of Automobile Engineering, Kongu Engineering College, Perundurai 638060, India
e-mail: priyankabhaskaran1993@gmail.com

S. Thangavel · B. Meenakshipriya · D. V. Prabu
Department of Mechatronics Engineering, Kongu Engineering College, Perundurai 638060, India
e-mail: thangavel.mts@kongu.ac.in

B. Meenakshipriya
e-mail: bmp@kongu.ac.in

D. V. Prabu
e-mail: venkatprabu13@gmail.com

N. S. Sivakumar
Department of Mechatronics Engineering, Tishk International University-TIU-Erbil, Erbil, Iraq
e-mail: sivakumar.ns@tiu.edu.iq

K. R. Ahmed et al. (eds.), *Deep Learning and Big Data for Intelligent Transportation*, Studies in Computational Intelligence 945,
https://doi.org/10.1007/978-3-030-65661-4_1

tasks at hand in Hadoop 1.x, they couldn't be applied to Hadoop 2.x because of basic building changes and dynamic asset assignment in Hadoop 2.x. Consequently, the proposed arrangement depends on a current presentation model for Hadoop 1.x, however thinking about building changes and catching the execution stream of a MapReduce work by utilizing lining network model. Thusly, the cost model mirrors the intra-work synchronization requirements that happen due the conflict at shared assets.

Keywords Big data · Hadoop · Challenges · Analytics methods

1 Introduction

These days, huge information volumes are every day produced at a phenomenal rate from heterogeneous sources (e.g., wellbeing, government, informal organizations, advertising, money related) [1]. This is because of numerous mechanical patterns, including the Internet of Things, the multiplication of Cloud Computing just as the spread of brilliant gadgets. Behind the scene, ground-breaking frameworks, and appropriated applications are supporting such various associations frameworks (e.g., intelligent network frameworks, medical care frameworks, retailing frameworks, government frameworks). Already to Big Data insurgency, organizations couldn't store all their files for extensive stretches nor effectively oversee gigantic information sets. Surely, conventional innovations have restricted capacity limit, unbending administration devices, and are costly [2]. They absence of versatility, adaptability and execution required in Big Data setting. Truth be told, Big Information the board requires noteworthy assets, new strategies and amazing advances. All the more accurately, Big Data needs to clean, measure, examine, make sure about, and give a granular admittance to enormous advancing informational indexes. Organizations and ventures are more mindful that information examination is progressively turning into a crucial factor in finding new knowledge and customizing administrations [3].

As a result of the intriguing worth that can be extricated from Big Information, numerous entertainers in various nations have dispatched significant ventures. In March 2012, the Official Data Administration dispatched Big Data Innovative Work Initiative with a financial plan of 200 million. Big Data improvement became one significant hatchet of the public innovative system in July 2012 and gave a report entitled Big Data for Development: Opportunities and Challenges [4]. It expects to layout the fundamental worries about Large Data challenges and to cultivate the discourse about how Big Data can serve the worldwide turn of events. Because of the diverse Big Data ventures over the world, numerous Big Data models, structures and new innovations were made to give more stockpiling limit, equal preparing and real-time examination of various heterogeneous sources. What's more, new arrangements have been created to guarantee information protection and security. Contrasted with conventional advances, such arrangements offer greater adaptability, versatility and execution [5]. Moreover, the expense of most equipment stockpiling and preparing arrangements is consistently dropping because of the reasonable mechanical development. Fig.1 shows the versatile components of Big data.

To remove information from Big Data, different models, programs, virtual products, equipment types and advancements have been planned and proposed. They attempt to guarantee more precise and solid outcomes for Big Information applications [6].

Nonetheless, in such condition, it might be time expending and testing to pick among various advancements. Truth be told, numerous boundaries ought to be thought of: innovative similarity, sending unpredictability, cost, effectiveness, execution, unwavering quality, backing and security hazards [7]. There exist some Big Information studies in the writing however the majority of them will in general zero in on calculations and approaches used to handle Big Data as opposed to advancements. Fig. 2 shows the interconnection of components in the industry ecosystem associated with data analytics.

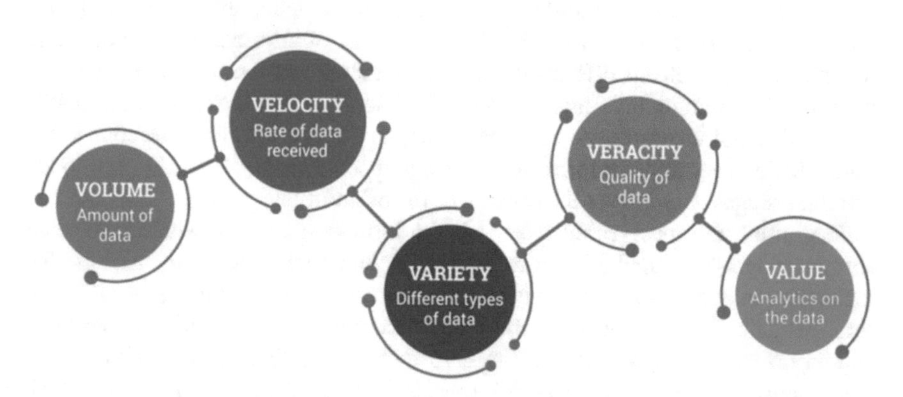

Fig.1 Versatile components of big data

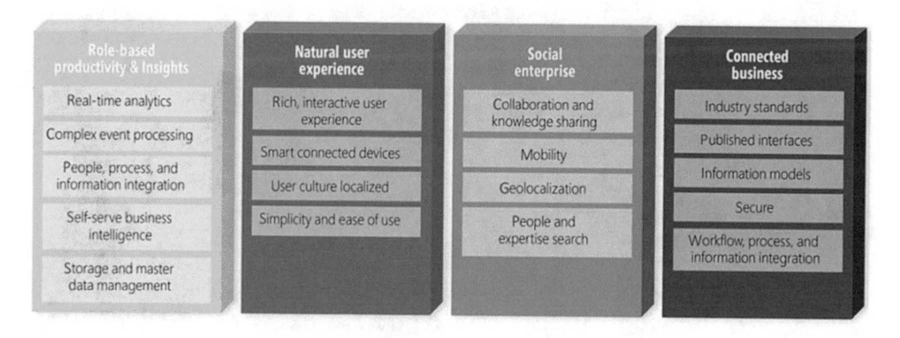

Fig. 2 Interconnection of Industry 4.0 ecosystem for data analytics

2 Relationship Between IoT and Big Data Analyitcs

Large information investigation is quickly developing as a key IoT activity to improve dynamic [8]. One of the most unmistakable highlights of IoT is its examination of data about "associated things." Big information examination in IoT requires preparing a huge measure of information on the fly and putting away the information in different capacity advances. Given that a significant part of the unstructured information are accumulated straightforwardly from web-empowered "things," enormous information executions will require performing lightning fast investigation with huge inquiries to permit associations to increase fast bits of knowledge, settle on snappy choices, and connect with individuals and different gadgets [9]. The interconnection of detecting what's more, activating gadgets give the ability to share data across stages through a brought together design and build up a typical working picture for empowering inventive applications.

The need to embrace large information in IoT applications is convincing. These two advances have just been perceived in its fields and business. In spite of the fact that, the advancement of enormous information is as of now slacking, these innovations are between subordinate and ought to be mutually evolved [10]. By and large, the arrangement of IoT builds the measure of information in amount and classification; henceforth, offering the chance for the application and improvement of enormous information investigation. Also, the use of large information innovations in IoT quickens the examination advances and plans of action of IoT. The connection among IoT and enormous information, which is appeared in Fig. 3, can be partitioned into three stages to empower the executives of IoT information [11].

The initial step involves overseeing IoT information sources, where associated sensors gadgets use applications to cooperate with each other. For instance, the connection of gadgets, for example, CCTV cameras, shrewd traffic lights, and keen home gadgets, creates a lot of information sources with various configurations [12]. This information can be put away in ease item stockpiling on the cloud. In the subsequent advance, the produced information is called "large information," which

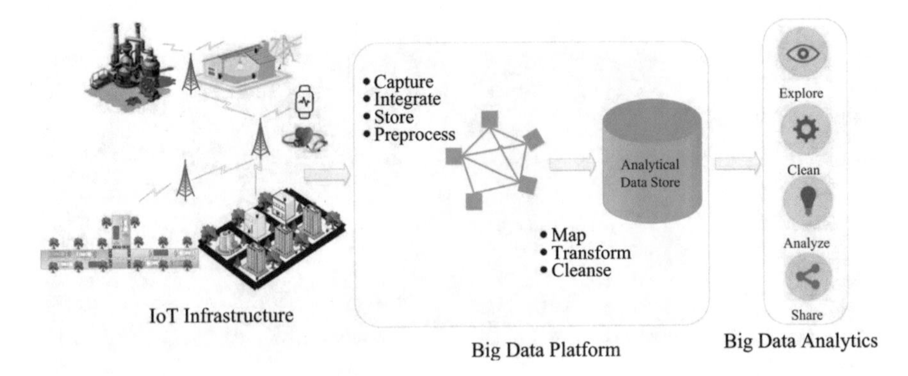

Fig. 3 Intercorrelation of IoT and big data

depend on their volume, speed, and assortment. These tremendous measures of information are put away in enormous information records in shared disseminated fault tolerant information bases. The last advance applies investigation devices such as MapReduce, Spark, Splunk, and Skytree that can investigate the put away huge IoT informational indexes. The four degrees of examination start from preparing information, at that point proceed onward to examination instruments, inquiries, also, reports [13].

MapReduce-based frameworks are progressively being utilized for enormous scope information examination applications. Limiting the execution time is imperative for MapReduce with respect to any information preparing application, and exact assessment of the execution time is fundamental for enhancing [14]. Hence, we have to manufacture execution models that follow the programming model of such information handling applications. Moreover, an away from of framework execution under various conditions is the way to basic choice making in remaining burden the board and asset scope quantification. Investigative execution models are especially appealing apparatuses as they may give sensibly exact occupation reaction time at altogether lower cost than exploratory assessment of genuine arrangements [15]. Programming in MapReduce requires adjusting a calculation to two-stage preparing model, i.e., Map and Reduce. Projects written in this practical style are consequently parallelized and executed on registering bunches. Apache Hadoop is one of the most mainstream open-source executions of MapReduce worldview. In the main form of Hadoop,1 the programming worldview of MapReduce and the asset the executives were firmly coupled [16]. So as to improve the general presentation just as the convenience and similarity with other appropriated information handling applications, a few prerequisites were included, for example, high bunch use, elevated level of unwavering quality and accessibility, uphold for programming model decent variety, and adaptable asset model. Along these lines, the design of the second form of Hadoop has gone through huge enhancements, presenting YARN (Yet Another Resource Negotiator), a different asset the executive module that discernibly changes the Hadoop engineering, which decouples the programming model from the asset the board framework and agents many booking capacities to per-application segments [17]. The bunch assets are currently being considered as nonstop, henceforth there is no static apportioning of assets (i.e., a division among plan and decrease openings). Hence, plan and lessen assignments contend now for similar assets. Obviously, it is difficult to apply the cost models transferring on such a static asset assignment as in the main form of Hadoop, and it is important to discover other approaches [18].

3 Big Data Analytics Methods

Large information examination plan to quickly extricate learned data that helps in making expectations, distinguishing late patterns, finding shrouded data, and at last, settling on choices. Information mining procedures are generally conveyed for both issue explicit techniques and summed up information investigation. In like manner, factual and AI strategies are used [19]. The development of enormous information moreover changes investigation prerequisites. Despite the fact that the necessities for effective components lie in all parts of large information the executives, for example, catching, stockpiling, preprocessing, and examination; for our conversation, large information investigation requires the same or quicker handling speed than customary information investigation with least expense for high-volume, high-speed, and high variety information [20]. Different arrangements are accessible for large information examination, and headways in creating and improving these arrangements are by and large consistently accomplished to make them appropriate for new enormous information patterns.

Information mining assumes a significant part in investigation, and the vast majority of the methods are created utilizing information mining calculations as indicated by a specific situation [21]. Information on accessible large information investigation alternatives is pivotal while assessing and picking a fitting methodology for dynamic. In this part, we present a few techniques that can be executed for a few major information contextual analyses. A portion of these investigation techniques are proficient for huge IoT information investigation. Different and enormous size informational indexes contribute more in large information experiences. Notwithstanding, this conviction isn't generally substantial in light of the fact that more information may have more ambiguities and variations from the norm [22]. We present large information investigation techniques under order, bunching, affiliation rule mining, and forecast classifications. Figure 4 portrays and sums up every one of these classifications.

Every classification is an information mining capacity and includes numerous strategies and calculations to satisfy data extraction also, investigation prerequisites.

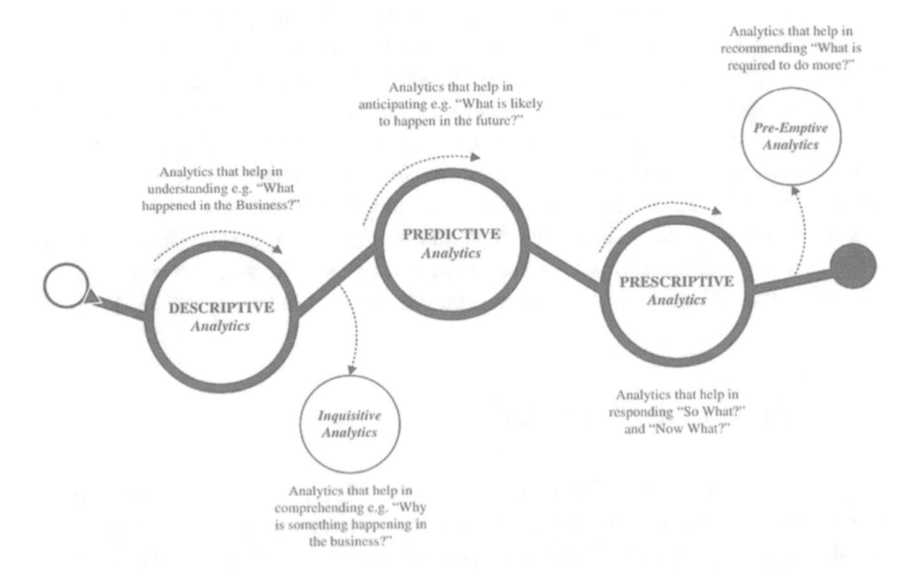

Fig. 4 Analytical methods associated with big data

For instance, Bayesian organization. Outline of large information investigation techniques. uphold vector machine (SVM), and k-closest neighbor (KNN) offer order techniques [23]. Correspondingly, parceling, progressive bunching, and co-event are broad in bunching. Affiliation rule mining and expectation include critical techniques. Order is a directed learning approach that employments earlier information as preparing information to order information objects into bunches. A predefined class is allotted to an item, and accordingly, the goal of anticipating a gathering or class for an item is accomplished. Discovering obscure or on the other hand concealed examples is all the more trying for huge IoT information [24]. Besides, removing important data from enormous information sets to improve dynamic is a basic errand.

A Bayesian network is a grouping strategy that offers model interpretability. Bayesian organizations are proficient for examining complex information structures uncovered through huge information rather than conventional organized information designs [25]. These organizations are coordinated non-cyclic charts, where hubs are irregular factors also, edges signify contingent reliance. Credulous, specific gullible, semi-guileless Bayes, and Bayes multi-nets are the proposed classes for grouping.

Bunching is another information mining strategy utilized as a major information investigation strategy [26]. As opposed to arrangement, grouping utilizes an unaided learning approach and makes gatherings for given articles dependent on their particular significant highlights. As we have introduced in Fig. 4 that gathering an enormous number of articles as bunches makes information control basic. The notable techniques utilized for bunching are progressive grouping and apportioning. The progressive bunching approach continues joining little bunches of information objects to frame a various leveled tree and make agglomerative groups. Disruptive groups are made in the inverse way by separating a solitary group that contains all information objects into littler proper groups. Market investigation and business dynamic are the hugest utilizations of huge information examination [27]. The cycle of affiliation rule mining includes distinguishing fascinating connections among various articles, occasions, or different substances to break down market patterns, purchaser purchasing conduct, furthermore, item request expectations. Affiliation rule mining centers around distinguishing and making rules in light of the recurrence of events for numeric and nonnumeric information. Information preparing is acted in two habits under affiliation rules [28]. To begin with, successive information preparing employments priori-based calculations, for example, MSPS and LAPINSPAM, to recognize cooperation affiliations. Another critical information preparing approach under affiliation rule is fleeting grouping examination, which utilizes calculations to break down occasion designs in consistent information [29]. Prescient examination utilizes recorded information, which are known as preparing information, to decide the outcomes as patterns or conduct in information. SVM and fluffy rationale calculations are utilized to distinguish connections among free and ward factors furthermore, to acquire relapse bends for forecasts, for example, for catastrophic events [30].

4 Big Data Challenges and Future Prespectives

The mining of Big Data offers numerous appealing chances. In any case, scientists and experts are confronting a few difficulties while investigating Big Data sets and when extricating esteem also, information from such mines of data. The challenges lye at various levels including: information catch, stockpiling, looking, sharing, investigation, the board and perception [31]. Besides, there are security and protection issues particularly in conveyed information driven applications. Frequently, the downpour of data and circulated streams outperform our capacity to outfit. Actually, while the size of Big Data continues expanding exponentially, the current mechanical ability to deal with and investigate Big Data sets, is as it were in the generally lower levels of petabytes, exabytes and zettabytes of information [32]. In this segment, we talk about in more subtleties some innovative issues actually opened for research. Figure 5 discloses the data-cycle of the challenges incorporated with Big Data.

Large Data the executives: Information researchers are confronting numerous difficulties when managing enormous data. One test is the way to gather, incorporate and store, with less equipment and programming prerequisites, enormous informational indexes produced from disseminated sources [33]. Another test is Big Data the board. It is vital to effectively oversee Big Data so as to encourage the extraction of solid knowledge and to improve costs. In reality, a decent information the board is the establishment for Big Data examination. Large Data the board intends to clean information for dependability, to total information originating from various sources and to encode information for security and protection. It implies likewise to guarantee proficient Big Information stockpiling and a job-based admittance to different dispersed endpoints [34]. At the end of the day, Big Data the executive's objective is to guarantee solid information that is effectively available, reasonable, appropriately put away and made sure about.

Enormous Data cleaning: Those five stages (Cleaning, Aggregation, Encoding, Storage and Access) are not new and are known on account of conventional information the executives. The test in Big Data is the manner by which to deal with the unpredictability of Big Data nature (speed, volume and assortment) and measure it

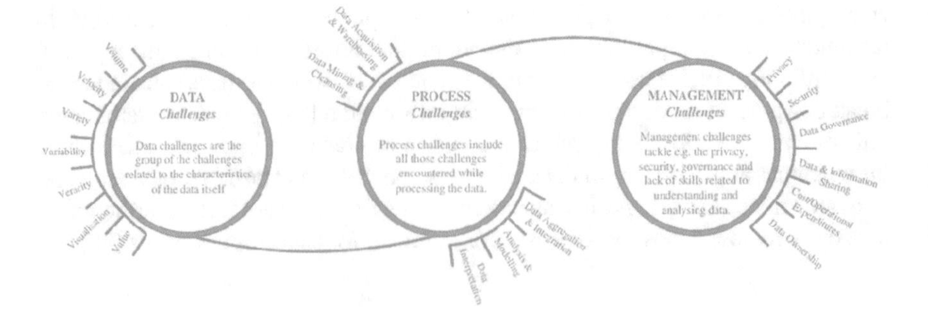

Fig. 5 Challenges incorporated with big data analytics

in a conveyed situation with a blend of uses. Indeed, for dependable examination results, it is basic to confirm the dependability of sources and information quality previously drawing in assets [35]. Be that as it may, information sources may contain clamors, mistakes or deficient information. The test is the way to clean such tremendous informational collections and how to choose about which information is solid, which information is valuable [36].

Large Data conglomeration: Another test is to synchronize outside information sources and circulated Big Data platforms (counting applications, vaults, sensors, organizations, and so on.) with the inner frameworks of an association. More often than not, it isn't adequate to break down the information created inside associations [37]. So as to remove significant understanding and information, it is imperative to go above and beyond also, to total interior information with outer information sources. Outside information could incorporate outsider sources, data about market change, climate determining and traffic conditions, information from interpersonal organizations, clients remark and resident criticisms [38]. This can help, for example, to expand the quality of prescient models utilized for analytics.

Imbalanced frameworks limits: A significant issue is identified with the PC engineering and limit. In fact, it is realized that the CPU execution is multiplying every year and a half observing the Moore's Law, and the exhibition of plate drives is likewise multiplying at a similar rate. Be that as it may, the I/O activities don't follow a similar presentation design [39]. (i.e., irregular I/O speeds have improved respectably while successive I/O speeds increment with thickness gradually). Subsequently, this imbalanced framework limits may slow getting to information and influences the presentation and the versatility of Big Data applications. From another edge, we can see the different gadgets limits over an organization (i.e., sensors, plates, recollections). This may hinder framework execution [40].

Imbalanced Big Data: Another test is arranging imbalanced dataset. This issue has picked up bunches of consideration in the most recent years. Truth be told, genuine world applications may create classes with various disseminations [41]. The primary kind of class that are under-given immaterial number of occasions (known as the minority or positive class). The inferior that have a bountiful number of occasions (known as the dominant part or negative class). Recognizing the minority classes is significant in different fields, for example, clinical conclusion, programming surrenders identification, Funds or bioinformatics [42]. The old-style learning procedures are not adjusted to imbalanced informational collections [43]. This is on the grounds that the model development depends on worldwide search measures without thinking about the quantity of examples [44].

In fact, worldwide principles are generally favored rather than explicit guideline so the minority class are ignored during the model development. Consequently, Standard learning strategies don't think about the distinction between the quantity of tests having a place with various classes [45]. Nonetheless, the classes which are under-spoken to may establish significant cases to distinguish. By and by, numerous difficult areas have multiple classes with lopsided disseminations, for example, protein overlay order and weld defect characterization. These multi-class awkwardness issues present new difficulties that are not watched in two-class issues [46]. Actually, managing multi-class undertakings with distinctive misclassification expenses of classes is more earnestly than managing with two-class ones. To tackle this issue, various techniques have been formed and frequently arranged into two classifications. The first one stretches out some parallel grouping procedures to make them material for multi-class arrangement issues, e.g., discriminant investigation, choice trees, k-closest neighbors, Naive Bayes, neural organizations, and backing vector machines [47]. The subsequent class is known as Decomposition and Ensemble Methods (DEM). It comprises of decaying a multi-class arrangement issue into a lot of double arrangement issues that can be understood by Binary Classifiers (BCs), and afterward characterizing a groundbreaking perception by applying an aggregative technique on the BCs' expectations [48].

5 Mapreduce Model for Dynamic Modeling Towards Scheduled Computing

The primary test in creating cost models for MapReduce occupations is that they should catch, with sensible precision, the different wellsprings of postpones that work may insight [49]. Specifically, assignments having a place with work may encounter two kinds of postponements: lining delays because of the dispute at shared assets, and synchronization delays because of the priority imperatives among assignments that coordinate in a similar activity—plan and diminish stages. There are two principle methods to assess the exhibition of remaining burdens of equal applications that locally don't consider the synchronization delays. One such procedure is Mean Value Analysis (MVA), which thinks about just undertaking queueing delays because of sharing of basic assets [50]. Along these lines, it can't be legitimately applied to remaining tasks at hand that have priority requirements, such as the synchronization among plan and decrease errands having a place with a similar MapReduce work. Elective traditional arrangement is to together abuse Markov Chains for speaking to the potential conditions of the framework, and lining network models, to process the change rates between states [51]. Nonetheless, such methodologies don't scale well since the state space develops exponentially with the quantity of errands, making it difficult to be applied to display occupations with numerous assignments, which is generally the situation of MapReduce work.

As per the model approval results, the proposed model produces evaluations of normal occupation reaction time that go astray from estimations of a genuine execution by under 15%. Despite the fact that this model doesn't catch the dynamic asset designation and it expect a fixed measure of strings to handle map also, decrease undertakings per hub as one of the info boundaries [52], it has significant preferences in correlation with past models. First of all, dissimilar to Herodotous' model that doesn't catch asset dispute between errands, this model is considering the lining delays because of the conflict at shared assets [53]. Besides, it is ready to catch the synchronization delays presented by the correspondence among plan and lessen assignments (ARIA and Tetris are not considering this property of MapReduce work execution) [54]. Figure 6 shows the task execution in hadoop by resource management of YARN.

In the second form of Hadoop, the YARN module showed up also, changed the design fundamentally. It is liable for overseeing bunch assets and occupation planning. In the past renditions of Hadoop, this usefulness was coordinated in the MapReduce module, where it was acknowledged by the JobTracker2 part. The JobTracker was answerable for booking, asset the board, checking and re-execution of bombed assignments, revealing position status to clients, recording review logs, collection of insights, client validation, and numerous others capacities [55]. The incredible measure of obligations caused restriction of adaptability. The basic thought of YARN is to part the two significant functionalities of the Job Tracker, asset the board and undertaking booking/checking so as to have a worldwide Resource Manager, furthermore, application-explicit Application Master. By isolating asset the executives capacities from the programming model, YARN delegates many planning

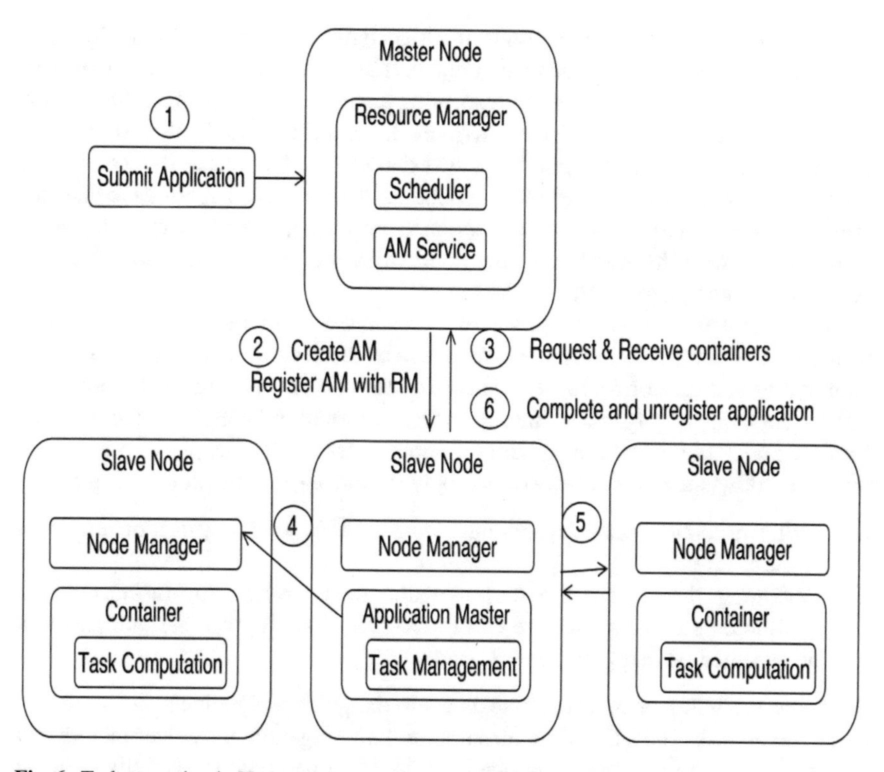

Fig. 6 Task execution in Hadoop in association with YARN

related assignments to per-work parts and totally withdraws from the static apportioning of assets for maps and lessens, considering the bunch assets as a continuum, which carries noteworthy enhancements to group use [56].

YARN comprises of three principle parts:

- Global Resource Manager (RM) per group, executing on the ace hub.
- Application Master (AM) per work.
- Node Manager (NM) per slave hub.

RM runs as a daemon on a committed machine and parleys all the accessible assets among different contending applications. We won't go in detail of all segments of RM3 and will zero in on the most significant ones:

- Scheduler, which is answerable for dispensing assets to the different applications that are running [57].
- Application Manager Service that arranges the primary holder (coherent heap of assets bound to a specific hub) for the Application Master. AMs are answerable for arranging assets with the RM and for working with the NMs to begin, screen, and stop the compartments [58].

In light of the center functionalities of YARN parts, the overall pattern of occupation execution measure is introduced in Fig. 6. It begins at the point when a customer presents a solicitation to the RM for executing an application (1). The AM registers with the RM through AM Service furthermore, is begun in the holder that AM Service committed for it (2). At that point, the AM demands holders from the RM to play out the genuine work (3). When the AM acquires holders, it can continue to dispatch them by conveying to a NM (4). Calculation takes place in the compartments, which stay in touch with the AM (5). At the point when the application is complete, the AM ought to unregister from the RM [59–62].

Instatement movement comprises of two sub exercises that can run in equal: instating the normal living arrangement season of each sort of task at each assistance community, and the normal reaction season of each task in the framework. For introducing the living arrangement time, we take the normal of habitation time from the historical backdrop of relating genuine Hadoop work executions [63]. To introduce the undertakings reaction time, we can apply the accompanying methodologies:

(a) Using inspecting methods—taking the normal of undertaking reaction time from work profile.
(b) Obtaining them from the current static cost models, for instance, from Herodotou's cost models (we can expect that first all guide assignments will be executed, at that point lessen errands).

Hence, we will give all accessible assets to the guide assignments and afterward to the lessen tasks [64]. Based on the structural investigation, the center factors that impact the timetable development measure are identified with the activity booking and to the asset the executive's framework, and can be characterized as following:

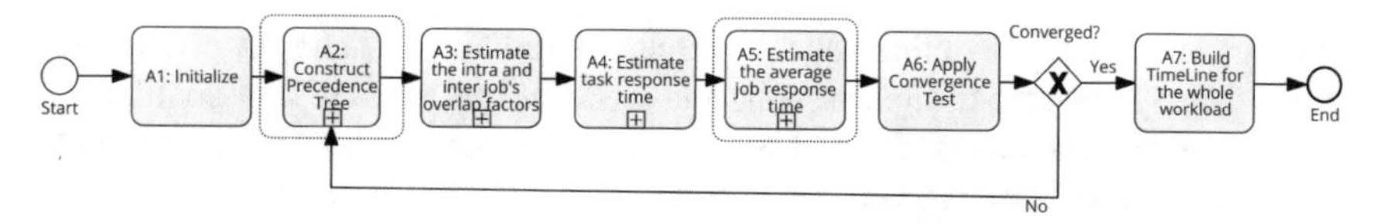

Fig.7 Data flow in accordance with processing, managing and analyzing task with Hadoop

1. RM Scheduler. We accept that RM utilizes the Capacity scheduler, which is the default scheduler of the Hadoop YARN circulation. The major unit of the Capacity scheduler is a line. Without loss of consensus, we accept a solitary, root line, in this way, asset allotment among occupations will be in the FIFO request (i.e., the need will be given to the principal work mentioning the assets).
2. AM Scheduler. Because of building changes, a few duties of occupation planning are designated to the AM, accordingly we need to decide the manner in which the AM appropriates holders for errands among various hubs. As indicated by discoveries in Fig. 7, plan and diminish errands have diverse lifecycles that we have to consider during the course of events development strategy.
3. Authoritative of assets. We are accepting that AM will utilize mentioned compartments for similar sort of undertakings as initially mentioned, in this way we overlook the late restricting usefulness of AM.

For assessing the activity reaction time, we have to know the conditions between errands (the priority tree reflects them) and the assessments for the errand terms. At that point, going from lower part of the priority tree to the top, we are assessing the spans for inside successive and equal hubs. At long last, the assessment of term for the root hub will be the length for the entire activity. As we continually going from base to the top, the profundity of priority tree effects on the blunder of employment reaction time assessment. So as to decrease the mistake, we balance the priority tree.

6 Conclusions

Current Big Data stages are upheld by different handling, investigative devices just as unique perception. Such stages empower to extricate information and incentive from complex dynamic condition. They likewise uphold dynamic through proposals and programmed identification of irregularities, unusual conduct or new patterns. In this chapter, we have concentrated Big Data qualities and profoundly examined the difficulties raised by Big Data registering frameworks. Not with standing that, we have clarified the estimation of Big Data mining in a few areas. Plus, we have zeroed in on the parts furthermore, advancements utilized in each layer of Big Data stages. Extraordinary advancements and disseminations have been additionally thought about in wording of their capacities, focal points and cutoff points. We have additionally arranged Big Data frameworks dependent on their highlights and administrations gave to definite clients. Along these lines, this chapter gives a definite knowledge into the design, systems and practices that are as of now continued in Big Data registering. Regardless of the significant advancements in Big Data field, we can see through our examination of different innovations, that many weaknesses exist. The vast majority of the time, they are identified with embraced structures and procedures.

The displaying approach broadens the arrangement proposed for Hadoop 1.x, where the execution stream of work was introduced by a priority tree and the conflict at the physical assets were caught by a shut lining organization. Our fundamental commitments are the profound examination of the Hadoop 2.x internals, distinguishing the primary building changes in Hadoop, and the making of the MapReduce execution model for Hadoop 2.x. Specifically, considering the distinguished changes in the design of Hadoop 2.x and considering the dynamic asset assignment, we made the technique for course of events development, in view of which the priority tree is assembled.

Acknowledgements This research work is carried out under the Senior Research fellowship received from CSIR (Council for Scientific and Industrial Research) with grant no.678/08(0001)2k18 EMR-I.

Conflict of Interest All the contributors to this research work have no clashes of attention to announce and broadcasting this article.

References

1. Babar M, Arif F, Jan MA, Tan Z, Khan F (2019) Urban data management system: towards big data analytics for internet of things based smart urban environment using customized Hadoop. Future Gener Comput Syst 96: 398–409
2. More PD, Nandgave S, Kadam M (2020) Weather data analytics using Hadoop with map-reduce. In: ICCCE. Springer, Singapore, pp 189–196
3. Babar M, Iqbal W, Kaleem S (2019) Internet of things based smart community design and planning using Hadoop-based big data analytics. In: Future of information and communication conference. Springer, Cham, pp 1046–1057
4. Lu J, Chen Y, Herodotou H, Babu S (2019) Speedup your analytics: automatic parameter tuning for databases and big data systems. Proc VLDB Endowment 12(12):1970–1973
5. Amalina F, Hashem IAT, Azizul ZH, Fong AT, Firdaus A, Imran M, Anuar NB (2019) Blending big data analytics: review on challenges and a recent study. IEEE Access 8:3629–3645
6. Astrova I, Koschel A, Heine F, Kalja A (2019) Moving Hadoop to the cloud for big data analytics. In: Databases and information systems X: selected papers from the thirteenth international baltic conference, DB&IS 2018, vol 315. IOS Press, p 195
7. Tariq H, Al-Sahaf H, Welch I (2019) Modelling and prediction of resource utilization of hadoop clusters: a machine learning approach. In: Proceedings of the 12th IEEE/ACM international conference on utility and cloud computing, pp 93–100
8. Guleria P, Sood M (2019) Big data analytics: educational data classification using Hadoop-inspired mapreduce framework. In: Predictive intelligence using big data and the internet of things. IGI Global, pp 77–108

9. Nakagami M, Fortes JAB, Yamaguchi S (2019) Job-aware optimization of file placement in hadoop. In: 2019 IEEE 43rd annual computer software and applications conference (COMPSAC), vol 2. IEEE, pp 664–669
10. Prabhu CSR, Chivukula AS, Mogadala A, Ghosh R, Livingston LMJ (2019) Big data tools—Hadoop ecosystem, spark and NoSQL databases. In: Big data analytics: systems, algorithms, applications. Springer, Singapore, pp 83–165
11. Basha SAK, Basha SM, Vincent DJ, Rajput DS (2019) Challenges in storing and processing big data using Hadoop and Spark. In: Deep learning and parallel computing environment for bioengineering systems. Academic Press, pp 179–187
12. Tahsin A, Hasan MM (2020) Big data and data science: a descriptive research on big data evolution and a proposed combined platform by integrating R and python on Hadoop for big data analytics and visualization. In: Proceedings of the international conference on computing advancements, pp 1–2 (2020)
13. Asaad RR, Ahmad HB, Ali RI (2020) A review: big data technologies with hadoop distributed filesystem and implementing M/R. Acad J Nawroz Univ 9(1):25–33
14. Sharma A, Singh G (2020) A review of scheduling algorithms in Hadoop. In: Proceedings of ICRIC 2019. Springer, Cham, pp 125–135
15. Jiang Y, Liu Qi, Dannah W, Jin D, Liu X, Sun M (2020) An optimized resource scheduling strategy for Hadoop speculative execution based on non-cooperative game schemes. Comput Mater Continua 62(2):713–729
16. Javanmardi AK, Hadi Yaghoubyan S, Bagherifard K, Nejatian S, Parvin H (2020) A unit-based, cost-efficient scheduler for heterogeneous Hadoop systems. J Supercomput:1–22
17. Upadhyay U, Sikka G (2020) STDADS: an efficient slow task detection algorithm for deadline schedulers. Big Data 8(1):62–69
18. Liang Y, Tang Y, Zhu X, Guo X, Wu C, Lin D (2020) Task scheduling strategy for heterogeneous spark clusters. In: Artificial intelligence in China. Springer, Singapore, pp 131–138
19. Abimbola M, Khan F, Khakzad N, Butt S (2015) Safety and risk analysis of managed pressure drilling operation using Bayesian network. Saf Sci 76:133–144
20. Aljaroudi A, Khan F, Akinturk A, Haddara M, Thodi P (2015b) Probability of detection and false detection for subsea leak detection systems: model and analysis. J Failure Anal Prevent
21. Kawsar MRU, Youssef SA, Faisal M, Kumar A, Seo JK, Paik JK (2015) Assessment of dropped object risk on corroded subsea pipeline. Ocean Eng 106:329–340
22. Khakzad N, Khan F, Amyotte P (2013) Dynamic safety analysis of process systems by mapping bow-tie into Bayesian network. Process Saf Environ Prot 91(1–2):46–53
23. Rausan M (2013) Risk assessment: theory, methods, and applications, Wiley
24. Fang N, Chen G, Zhu H, Meng H (2014) Statistical analysis of leakage accidents of submarine pipeline. Oil Gas Storage Transp 33(01):99–103
25. Zhang X, Xie L, Chen G (2011) Integrity management technique for submarine pipeline. Oil Field Equipm 40(12):10–15
26. Priyanka EB, Maheswari C (2016) Parameter monitoring and control during petrol transportation using PLC based PID controller. J Appl Res Technol 14(5):125–131
27. Priyanka EB, Maheswari C, Thangavel S (2018) Remote monitoring and control of an oil pipeline transportation system using a fuzzy-PID controller. Flow Meas Instrum 62(3):144–151
28. Priyanka EB, Maheswari C, Thangavel S (2018) IoT based field parameters monitoring and control in press shop assembly. Internet Things 3:1–11
29. Priyanka EB, Maheswari C, Thangavel S (2019) Remote monitoring and control of LQR-PI controller parameters for an oil pipeline transport system. Proc Inst Mech Eng Part I J Syst Control Eng 233(6):597–608
30. Priyanka EB, Maheswari C, Thangavel S (29018) Proactive decision making based IoT framework for an oil pipeline transportation system. In: International conference on computer networks, big data and IoT, 19 Dec 2018. Springer, Cham, pp 108–119
31. Priyanka EB, Krishnamurthy K, Maheswari C (2016, Nov) Remote monitoring and control of pressure and flow in oil pipelines transport system using PLC based controller. In: 2016 online international conference on green engineering and technologies (IC-GET). IEEE, pp 1–6

32. Subramaniam T, Bhaskaran P (2019) Local intelligence for remote surveillance and control of flow in fluid transportation system. Adv Modell Anal C 74(1):15–21. https://doi.org/10.18280/ama_c.740102
33. Atzori L, Iera A, Morabito G, Nitti M (2012) The social internet of things (SIot)—when social networks meet the internet of things: concept, architecture and network characterization. Comput Netw 56(16):3594–3608
34. Pavan Kumar YV, Bhimasingu R (2015) Key aspects of smart grid design for distribution system automation: architecture and responsibilities. Proc Technol 21(9):352–359. ISSN 2212-0173
35. Meribout M (2011) A wireless sensor network based infrastructure for real-time and online pipeline inspection. IEEE Sens J 11(11):2966–2972
36. Priyanka E, Maheswari C, Ponnibala M, Thangavel S (2019) SCADA based remote monitoring and control of pressure and flow in fluid transport system using IMC-PID controller. Adv Syst Sci Appl 19(3):140–162. https://doi.org/10.25728/assa.2019.19.3.676
37. Priyanka EB, Maheswari C, Thangavel S, Bala MP (2020) Integrating IoT with LQR-PID controller for online surveillance and control of flow and pressure in fluid transportation system. J Indust Inf Integr 17:100127. https://doi.org/10.1016/j.jii.2020.100127
38. Priyanka EB, Maheswari C, Thangavel S A smart-integrated IoT module for intelligent transportation in oil industry. Int J Numer Modell Electron Netw Dev Fields:e2731. https://doi.org/10.1002/jnm.2731
39. Maheswari C, Priyanka EB, Thangavel S, Vignesh SR, Poongodi C (2020) Multiple regression analysis for the prediction of extraction efficiency in mining industry with industrial IoT. Prod Eng Res Devel 14(4):457–471. https://doi.org/10.1007/s11740-020-00970-z
40. Bhaskaran PE, Chennippan M, Subramaniam T (2020) Future prediction and estimation of faults occurrences in oil pipelines by using data clustering with time series forecasting. J Loss Prev Process Ind 66:104203. https://doi.org/10.1016/j.jlp.2020.104203
41. Priyanka EB, Thangavel S and Gao, XZ (2020) Review analysis on cloud computing based smart grid technology in the oil pipeline sensor network system. Petroleum Research. https://doi.org/10.1016/j.ptlrs.2020.10.001
42. Priyanka EB, Thangavel S, Madhuvishal V, Tharun S, Raagul KV, Shiv Krishnan CS Application of integrated IoT framework to water pipeline transportation system in smart cities. In: Intelligence in big data technologies—beyond the hype. Springer, Singapore, pp 571–579. https://doi.org/10.1007/978-981-15-5285-4_57
43. Priyanka EB, Thangavel S, Pratheep VG (2020) Enhanced digital synthesized phase locked loop with high frequency compensation and clock generation. Sens Imag 21(1):1–12. https://doi.org/10.1007/s11220-020-00308-0
44. Maheswari C, Priyanka EB, Thangavel S, Parameswari P (2018) Development of unmanned guided vehicle for material handling automation for industry 4.0. Int J Recent Technol Eng 7(4s):428–432
45. Priyanka EB, Thangavel S, Parameswari P (2019) Automated pay and use browsing and printing machine. Int J Innov Technol Explor Eng (IJITEE) 8(11S):148–152
46. Pratheep VG, Priyanka EB, Raja R (2019) "Design and fabrication of 3-axis welding robot. Int J Innov Technol Explor Eng (IJITEE) 8(11):1588–1592
47. Priyanka EB, Thangavel S, Parameswari P (2019) Collision waring system using RFID in automotives. Int J Innov Technol Explor Eng (IJITEE) 8(11S):153–158
48. Maheswari C, Priyanka EB, Ibrahim Sherif IA, Thangavel S, Ramani G (2020) Vibration signals-based bearing defects identification through online monitoring using LABVIEW. J Eur Des Systèmes Automatisés 53(2):187–193
49. Priyanka EB, Thangavel S (2020) Influence of internet of things (IoT) in association of data mining towards the development smart cities-A review analysis. J Eng Sci Technol Rev 13(4):1–21. https://doi.org/10.25103/jestr.134.01
50. Wang X, Wang Y, Cui Y (2014) A new multi-objective bi-level programming model for energy and locality aware multi-job scheduling in cloud computing. Future Gener Comput Syst 36(7):91–101

51. Balaji P, Zeadally S, Malluhi QM, Tziritas N, Vishnu A, Khan SU, Zomaya A (2016) A survey and taxonomy on energy efficient resource allocation techniques for cloud computing systems. Computing 98(7):751–774
52. Chang V, Walters RJ, Wills G (2012) Cloud storage and bioinformatics in a private cloud deployment: lessons for data intensive research. In: International Conference on Cloud Computing and Services Science, vol 367, Springer International Publishing, pp 245–264
53. O'Driscoll A, Daugelaite J, Sleator RD (2013) 'Big Data', hadoop and cloud computing in genomics. J Biomed Inform 46(5):774–781
54. Merlo A, Clematis A, Corana A, Gianuzzi V (2011) Quality of service on grid: architectural and methodological issues. Concurr Comput 23(8):745–766
55. Bhaskaran PE, Maheswari C, Thangavel S, Ponnibala M, Kalavathidevi T, Sivakumar NS (2021) IoT based monitoring and control of fluid transportation using machine learning. Comput Electr Eng 89, p.106899. https://doi.org/10.1016/j.compeleceng.2020.106899
56. Pletea D, Pop F, Cristea V (2012) Speculative genetic scheduling method for Hadoop environments. In: 2012 14th international symposium on symbolic and numeric algorithms for scientific computing (SYNASC), IEEE, pp 281–286
57. Polo J, Carrera D, Becerra Y, Steinder M, Whalley I (2010) Performance driven task co-scheduling for MapReduce environments. In: Network operations and management symposium (NOMS). IEEE, pp 373–380
58. Pop F, Dobre C, Cristea V (2008) Performance analysis of grid dag scheduling algorithms using MONARC simulation tool. In: Parallel and distributed computing, 2008. International symposium on ISPDC'08. IEEE, pp 131–138
59. Prodan R, Sperk M (2013) Scientific computing with Google app engine. Future Gener Comput Syst 29(7):1851–1859
60. Sahni J, Vidyarthi DP (2016) Workflow-and-platform aware task clustering for scientific workflow execution in cloud environment. Future Gener Comput Syst 64:61–74
61. Sandholm T, Lai K (2010) Dynamic proportional share scheduling in Hadoop. In: Job scheduling strategies for parallel processing. Springer, pp 110–131
62. Maheswari C, Bhaskaran PE, Subramaniam T, Meenakshipriya B, Krishnamurthy K, Kumar VA (2020) Design and experimental investigations on NOx emission control using FOCDM (Fractional-Order-Based Coefficient Diagram Method)-PIλDμ Controller. J Européen des Systèmes Automatisés 53(5):695–703. https://doi.org/10.18280/jesa.530512
63. Priyanka EB, Thangavel S, Prabu (2020) Fundamentals of wireless sensor networks using machine learning approaches: advancement in big data analysis using hadoop for oil pipeline system with scheduling algorithm. Deep Learning Strategies for Security Enhancement in Wireless Sensor Networks. IGI Global, 233–254. https://doi.org/10.4018/978-1-7998-5068-7.ch012
64. Serwadda A, Phoha VV (2015) When mice devour the elephants: a DDoS attack against size-based scheduling schemes in the internet. Comput Secur 53:31–43

Big Data for Autonomous Vehicles

Rinki Sharma

Abstract Autonomous Vehicles (AVs) are going to be an integral part of the Intelligent Transportation Systems (ITS) in the future. To make AV a reality, it is essential to make the vehicles capable of communication with their surroundings and other vehicles. This gives rise to the Vehicle-to-Everything (V2X) communication scenarios. Realization of V2X communication necessitates real-time intra and inter vehicle communication. Such communication involves high data rate, and leads to enormous data required to be processed in real-time. The big data technologies play a significant role in making AV a reality. This chapter emphasizes the role of big data technologies in AVs. This chapter discusses the enabling technologies and communication protocols for AVs. The components involved in different AV applications and rate at which the data is generated for these applications is presented. The role of different big data aspects such as data acquisition, processing, storage, analysis, computing, transmission and security, are discussed by virtue of related research in these domains. With big data for AV being an upcoming research area, the open research issues, important research avenues and ongoing research in these areas is discussed is this chapter with the aim to help the researchers interested in pursuing research in this field.

Keywords Autonomous vehicles (AV) · Big data · Vehicular communication · Intelligent transportation systems (ITS) · Vehicle-to-everything (V2X)

Abbreviations

5GAA	5G Automotive Association
ABS	Antilock Braking System
ADAS	Advanced Driver Assistance System
AEC	Australian Engineering Conference

R. Sharma (✉)
Ramaiah University of Applied Sciences, Bengaluru, India
e-mail: rinki.cs.et@msruas.ac.in

K. R. Ahmed et al. (eds.), *Deep Learning and Big Data for Intelligent Transportation*, Studies in Computational Intelligence 945,
https://doi.org/10.1007/978-3-030-65661-4_2

AECC	Automotive Edge Computing Consortium
AM	Amplitude Modulation
AV	Autonomous Vehicle
BSM	Basic Safety Message
BT	Bluetooth
CAN	Controller Area Network
CAN-FD	Controller Area Network with Flexible Data Rate
CCN	Content Centric Networking
CD	Compact Disc
C-V2X	Cellular-Vehicle to Everything
DDoS	Distributed Denial of Service
DoS	Denial of Service
DPDA	Dual Polarized Directional Antenna
DSRC	Dedicated Short-Range Communication
ECU	Electronic Control Unit
FM	Frequency Modulation
GPS	Global Positioning System
HVAC	Heat Ventilation and Air Conditioning
ICN	Information Centric Networking
IDS	Intrusion Detection Systems
IEEE	Institute of Electrical and Electronics Engineers
ITS	Intelligent Transportation System
IVN	In-Vehicle Network
JTAG	Joint Test Action Group
LDWS	Lane Departure Warning System
LIDAR	Light Detection and Ranging
LIN	Local Interconnect Network
M2M	Machine to Machine
MANET	Mobile Ad-hoc Network
MITM	Man in the Middle
ML	Machine Learning
MOST	Media Oriented System Transport
MU-MIMO-OFDM	Multi User–Multi Input Multi Output Orthogonal Frequency Division Multiplexing
NDN	Named Data Networking
NDVN	Named Data Vehicular Network
NFV	Network Function Virtualization
NHTSA	National Highway Traffic Safety Administration
OBD	On Board Diagnostics
OTA	Over the Air
QoS	Quality of Service
RAID	Redundant Array of Independent Disks
RDS	Radio Data System
RSU	Road Side Unit
SAE	Society of Automotive Engineers

SDN	Software Defined Networking
SDVN	Software Defined Vehicular Network
TCP	Transmission Control Protocol
TCU	Transmission Control Unit
UDP	User Datagram Protocol
USB	Universal Serial Bus
V2C	Vehicle to Cloud
V2D	Vehicle to Device
V2G	Vehicle to Grid
V2I	Vehicle to Infrastructure
V2P	Vehicle to Pedestrian
V2R	Vehicle to Roadside
V2V	Vehicle to Vehicle
V2X	Vehicle to Everything
VANET	Vehicular Ad-hoc Network
VLC	Visible Light Communication
V-NDN	Vehicular-Named Data Network
WAVE	Wireless Access for Vehicular Environments
WSMP	WAVE Short Message Protocol

1 Introduction

Over the years there has been tremendous advancement in vehicular technology. Network and communication technology made its way into the vehicles for applications such as comfort, driver assist and fleet management. Gradually vehicle communication is advancing towards Vehicle-to-Everything (V2X) scenario. Under V2X communication, a vehicle is capable of Vehicle-to-Vehicle (V2V), Vehicle-to-Infrastructure (V2I), Vehicle-to-Cloud (V2C), Vehicle-to-Pedestrian (V2P), Vehicle-to-Device (V2D) and Vehicle-to-Grid (V2G), to name a few. There are standards and protocols defined to enable such communication. Enabling vehicles to communicate with their surroundings eventually aims to achieve an Autonomous Vehicle (AV), capable of maneuvering itself depending on its surroundings [1]. Realizing an AV requires high rate Intra and Inter vehicle communication over different interfaces of the vehicle. This leads to generation of enormous data, that needs to be processed at a very high rate. This is where big data technology plays a significant role in AVs.

This chapter aims to emphasize upon the importance of big data technologies in AVs. AVs are going to be a reality in future, and it will lead to generation of enormous data. Technologies should be made capable of communicating, storing and processing such data efficiently, reliably and securely. This chapter presents the enabling communication technologies and protocols for AVs. There are different protocols and technologies for different applications. The protocols and standards for vehicular communication, along with their supporting data rates are discussed in detail. Based on the levels of automation of an AV, the complexity and generated data increases. The components involved for different AV applications and rate of data generation is presented. After getting the clarity on big data generation of AV, the aspects of big data such as big data acquisition, processing, storage, analysis, computing, transmission and security, are discussed by virtue of related research in these domains. As big data for AV is an upcoming research area, there are many open research issues in this field. Lastly, some of the important research avenues in this field are presented.

2 Vehicular Communication

Vehicular communication is an integral part of the future Intelligent Transportation Systems (ITS). ITS comprises of intelligent vehicles capable of communicating with their surroundings and other vehicles that are within communication range. Eventually this trend will lead to self-driving or AV [2]. Intelligent vehicle applications can be broadly classified into the following four major areas:

1. **Convenience**: Aimed to assist the driver with safe and comfortable driving, applications such as infotainment, night vision assist, seat and mirror control, 360° camera, nearby services (restaurants, gas station, emergency etc.) and navigation assist fall in this category.
2. **Safety**: Advanced Driver Assistance Systems (ADAS), blind spot assist, lane detection and cruise control, vehicle platooning, crash detection, pre-safe brake are some of the applications that fall in this category.
3. **Productivity**: This applies to commercial and transit vehicles. Applications such as fleet management wherein parameters such as location, speed, mileage, drive time, idle time, stop time, trip reports of fleet and transit vehicles is traced and tracked to ensure efficient and productive maneuvering.
4. **Traffic assist**: This is used for improved and smooth traffic flow in case of crash and congestion. Road crashes are ‘spot disasters’, while traffic congestion is a ‘distributed disaster’, both affecting the smooth flow of the traffic. V2V and V2I communication is significant in propagating the information about crashes and traffic congestion. Based on this information, alternate routes can be used and the traffic can be diverted to appropriate routes, thus achieving smooth flow of traffic.

Vehicular communication implies of communication considering vehicles or the Electronic Control Units (ECUs) within the vehicle as the communication entities. Broad classification of vehicular communication is presented in Fig. 1.

Vehicular communication can be broadly classified as:

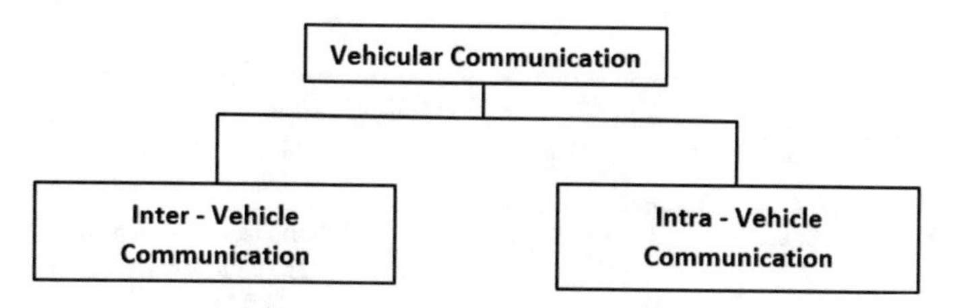

Fig. 1 Classification of vehicular communication

1. **Inter-vehicle communication**
   a. Primarily used for ITS applications such as convenience, safety, productivity and traffic assist. Inter-vehicle communication comprises of communication between vehicles as well as between vehicles and infrastructure. Some examples of Inter-vehicle communication are V2V, V2I, V2C, V2P, V2D and V2G, together represented as V2X communication [3]. By using inter-vehicle communication techniques, vehicles can communicate with each other for platooning/collision avoidance. A vehicle can communicate with roadside infrastructure, pedestrian or cloud for dynamic traffic signaling, transmission of safety alerts, real-time traffic and weather updates, road condition-based navigation and other cloud services. When two or more vehicles come within communication range of each other, they form an adhoc network and exchange data. Such networks are called as vehicular adhoc networks (VANETs). VANETs are a subset of mobile adhoc networks (MANETs), wherein the mobile nodes are vehicles having a pre-determined and uniform mobility pattern (unlike random mobility of MANETs) [4].
2. **Intra-vehicle communication**
   a. Intra-vehicle communication is also known as in-vehicle communication. It takes place within the vehicle, between its ECUs, sensors and actuators. Depending upon the application and required data rate, different in-vehicle networking protocols such as Controller Area Network (CAN), Local Interconnect Network (LIN), Media Oriented System Transport (MOST), FlexRay and Automotive Ethernet are used for communication between the components within the vehicle [5]. Details of in-vehicle networking protocols, their applications and required data rates are presented in Table 1.

3 Protocols and Standards for Vehicular Communication:

With the aim to make high speed exchange of data possible for inter- and intra-vehicle communication scenarios, numerous standardization activities have taken place. Among the frontrunners to develop protocols and standards for vehicular communication are Institute of Electrical and Electronics Engineers (IEEE) [6],

Table 1 In-vehicle network protocols

Protocol	Supported data rate	Applications
Controller area network (CAN)	1 Mbps	Engine management, cruise control, transmission control, ABS, HVAC
CAN with flexible data rate (CAN-FD)	15 Mbps	Airbags, antilock braking system (ABS), X-by-wire, all applications handled by CAN
Local interconnect network (LIN)	20 kbps	Control of door, window, seat, mirror, sunroof. Diagnostics
Media oriented system transport (MOST)	25 Mbps	Infotainment, navigation
FlexRay	10 Mbps	X-by-wire, airbags, antilock braking system (ABS)
Automotive ethernet	10 Mbps–1 Gbps	Backbone network, 360° camera, systems for infotainment, powertrain, ADAS, body comfort, chassis safety and diagnostics

Society of Automotive Engineers (SAE) [7] and 5GAA (5G Automotive Association) [8]. Apart from the standards developed by these groups, technologies such as Bluetooth and Visible Light Communication (VLC) are also used for short range vehicular communication [9]. The in-vehicle network communication protocols enable communication among the components within the vehicle. This section presents the protocols and standards used for vehicular communication.

3.1 Dedicated Short-Range Communication (DSRC)

IEEE and SAE worked together to develop the Dedicated Short-Range Communication (DSRC) standard for vehicle-to-vehicle (V2V) and vehicle-to-roadside (V2R) also known as V2I communication. DSRC standard comprises of Wireless Access for Vehicular Environments (WAVE), IEEE 802.11p amendment for vehicular environments, the 1609.2, 1609.3, 1609.4 standards for resource management, security services, networking services and multi-channel operations respectively [10]. While the communication stack is developed by IEEE (being the pioneers in network communication), the messages to be exchanged based on the application are defined by SAE. The SAE J2735 standard presents the message set dictionary and SAE J2945.1 standard defines the minimum communication performance requirements [11]. The layered architecture for DSRC standard is presented in Fig. 2.

DSRC operates at the 5.9 GHz frequency band that spans from 5.850 to 5.925 GHz. The short to medium range communication can go up to 1000 m (1 km). The supported V2V and V2R communication environments can be used for applications such as broadcasting localized traffic or road information directly into the vehicle,

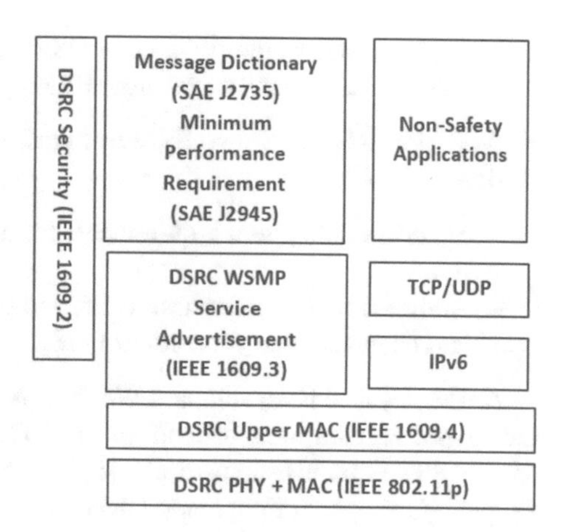

Fig. 2 Layered architecture of DSRC

highway advisory radio, emergency vehicle approach warning and impending collision warning, to name a few. The choice between using WAVE Short Message Protocol (WSMP) or IPv6+UDP/TCP depends on the requirements of a given application [12].

The protocols in the DSRC layered architecture and their functionalities are as follows:

- **IEEE 802.11p: Medium Access Control and Physical Layer Specifications for WAVE**
  - Amendment to 802.11 to enable operation without setting up a basic service set
  - Physical layer for implementing DSRC [13]
- **IEEE 1609.4: Multi-Channel Operations**
  - Coordinate switching between the control channel and service channels
  - User priority access to the media
  - Routing data packets on the correct channel [14]
- **IEEE 1609.3: Network Services**
  - Defines network and transport layer services, including addressing and routing
  - Examples of message types are WAVE short messages, WAVE service advertisements, and WAVE routing advertisements [15]
- **IEEE 1609.2: Security Services for Applications and Management Messages**
  - Defines secure message formats and processing
  - Messages to be protected from attacks such as eavesdropping, spoofing, alteration, and replay [16]
- **SAE J2735: DSRC Message Set Dictionary**
  - Specifies message sets, data frames, and data elements

  - Support interoperability among DSRC applications
  - Defines Basic Safety Message (BSM) [17]

- **SAE J2945/1: On-Board System Requirements for V2V Safety Communications**
  - Specifies the system requirements for an on-board (V2V) safety communications
  - Addresses BSM transmission rate and power, accuracy of BSM data elements, and channel congestion control [18].

As the physical layer standard WAVE is based on the 802.11 family of standards, the supported data rates depend on the 802.11 standards. The latest 802.11 standard is 802.11ax, also known as Wi-Fi 6 that uses Multi User-Multi Input Multi Output—Orthogonal Frequency Division Multiple Access (MU-MIMO-OFDMA). It can support data rates of up to 600 Mbps (80 MHz bandwidth, 1 spatial stream) and 9607.8 Mbps (160 MHz bandwidth, 8 spatial streams) [19].

3.2 Cellular V2X

Cellular V2X (C-V2X) standard is developed by 5GAA that works towards roadmap for 5G connectivity for vehicular communication. This standard has evolved from 3GPP standardized 4G LTE (Long Term Evolution) and 5G mobile cellular connectivity. C-V2X is considered to be capable of supporting highly reliable real-time communication at high speeds, in high-density traffic. Since this standard has evolved from 3GPP standards with established infrastructure all over the world, it reduces the cost of infrastructure deployment and benefits from cellular network densification. 3GPP support of ITS 5.9 GHz band. C-V2X and 802.11p can co-exist by being placed on different channels in the ITS band. The data rate and communication range are comparable with DSRC [20].

3.3 Bluetooth

The IEEE 802.15.1 standard, works in the unlicensed band of 2.4 GHz and supports 1 Mbps data rate for voice and data. It supports short range vehicular communication for up to 10 m distance. In vehicles, Bluetooth is used for wireless sensors and cable replacement [21]. Seat occupancy detection, mirror control, smart vehicle access (key fob control), personalization and infotainment control, assisted car parking and work zone indication through Bluetooth enabled barricades are some of the applications supported by Bluetooth in vehicular environments.

3.4 *Visible Light Communication (VLC)*

Supported by the IEEE 802.15.7 working group, VLC supports short range Line of Sight (LoS) communication. Visible light communication (VLC) uses visible light between 400 and 800 THz (780–375 nm). Fluorescent lamps can transmit signals at 10 kbps. LEDs can support data rate of up to 500 Mbps over short distances [22]. V2I communication is possible between traffic light and vehicle head lights while V2V communication is enabled through head light of a vehicle to tail lights of the preceding vehicle. Outdoor advertising on mobile nodes is another application of VLC.

3.5 *In-vehicle Communication Protocols*

A modern luxury car can have as many as 150 ECUs [23], that can handle complex and significant applications such as cruise control, engine management system, airbags, antilock braking system (ABS), heating ventilation and air conditioning (HVAC) that require high reliability and data rate, to simple tasks such as mirror control and seat adjustment. Infotainment systems in the car provide real-time data transmission that requires high data rates. Different protocols are used based on varying requirements of the applications. Table 1 presents the most popular wired protocols used for in-vehicle communication, their applications and data rates. As discussed in Sect. 3.3, Bluetooth wireless communication protocol is also used for in-vehicle communication.

4 Autonomous Vehicles (AVs)

An AV can be defined as the one that can drive itself from one point to another predetermined destination in 'autopilot' mode, using various in-vehicle and inter-vehicle communication technologies and applications, GPS navigation, sensors, lasers and radar. Use of AV is expected to rise considerably in coming years as they are expected to increase road safety and travel time efficiency.

Some of the advantages of using AVs are as follows:

1. **Road safety**: According to the study by National Highway Traffic Safety Administration (NHTSA), 94% of road accidents are caused by human error. Some of the main causes of human error while driving are over speeding, reckless driving, inattentiveness, intoxication, impaired and underage driving.
2. **Travel time efficiency**: Travel can be more productive for people when most of the driving decisions are automated. AVs use navigation and communicate with surrounding vehicles and infrastructure, which makes them capable of identifying road conditions, accidents or delays thus choosing better routes to destination.

3. **Reduced air and noise pollution**: Vehicular pollution is one of the main causes of increase in carbon monoxide and nitrogen oxide in the air. Slow moving traffic and congestion leads to high fuel burning leading to air pollution. Excessive honking or congested roads cause noise pollution. AVs would reduce this drastically through intelligent navigation and communication with surrounding vehicles and infrastructure.

4.1 Levels of Automation

SAE J3016 defines six levels of automation. The details of these levels are as follows [24]:

1. **Level 0 (L0)**: These are the basic cars without any form of automation.
2. **Level 1 (L1)**: Supports low level driver assistance systems. Vehicles with features such as lane departure assist and adaptive cruise control fall under this category. BMW's driver assistance systems are an example of L1 automation.
3. **Level 2 (L2)**: This level provides partial driving automation. Advanced Driver Assistance Systems (ADAS), wherein a vehicle can control both steering and acceleration/deceleration of the vehicle, fall under this category. Tesla's Autopilot and General Motor's Cadillac are examples of vehicles having L2 of automation.
4. **Level 3 (L3)**: This level supports conditional driving automation. L3 vehicles are capable of 'environmental detection' and make informed decisions. However, the driver is required to be alert and is allowed to take control of the system if necessary. Audi's A8 is a L3 vehicle.
5. **Level 4 (L4)**: These are high driving automation vehicles wherein the vehicle itself is capable to intervene in case of system failure, requiring no or minimal human interaction. While, L4 vehicles can operate in self-driving mode, there is still an option of manual control. Most of the L4 vehicles are being developed for taxi and ridesharing purposes. Some of the examples of L4 vehicles are shuttles and cabs by French company NAVYA, self-driving taxi by Waymo and Magna, and robotaxi by Volvo and Baidu.
6. **Level 5 (L5)**: These vehicles support full-driving automation and require no human control. These vehicles won't even have steering wheels or brakes, and will be free from geofencing. These vehicles are under test.

4.2 Big Data Sources for Autonomous Vehicle

AVs are capable of making informed decisions and self-drive because they constantly communicate with vehicles and infrastructure within their communication range, Global Positioning System (GPS), as well as the cloud [25].

The primary big data generation sources, related components and their applications are presented in Table 2.

Table 2 Big data generation sources

Technology/domain	Components	Applications
Imaging for automotive	Visible cameras 3D cameras Night vision cameras LIDAR Long-range radar Short-range radar Dead reckoning sensors Ultrasound GPS	Blind-spot Side-view (mirrorless cars) Stereo cameras: direction and distance for Lane Departure Warning System (LDWS) Traffic sign recognition Gesture recognition Presence detection Driver monitoring Pedestrian/animal detection 3D mapping of surroundings Adaptive cruise control Front and rear parking Odometry Parking Pedestrian and obstacle detection (short range) Positioning and navigation
Cloud	Data servers Roadside infrastructure	High definition maps Infotainment Vehicle sensor data Crowdsensing data Edge caching data Over The Air (OTA) updates Road condition updates
VANET communication	Wireless communication module	Cooperative driving Platooning Road condition updates (wet road, collision, braking, road maintenance) E-toll Location based advertising Retail promotion
In-vehicle network data	Electronic control unit (ECU) Sensors Actuators In-vehicle network	Engine control Tire pressure monitoring Transmission control Anti-lock braking Body control HVAC Infotainment X-by-wire Airbag Diagnostics ADAS Adaptive cruise control Seat and mirror control Backbone networks Gateways

According to a study at Australian Engineering Conference (AEC), 2018, an AV will generate around **166 GB of data per hour,** and around **4000 GB of data per day** [26]. Real-time applications such as traffic management, navigation, autonomous driving, toll collection, vehicular communication, review of vehicle performance and predictive maintenance, location-based promotion and advertising to name a few, require high data rate and bandwidth for acceptable operation and performance.

5 Big Data in AV

As seen from the examples of data generation in AVs, thousands of Giga Bytes of data per day will be produced and communicated in a network of AVs. Advanced, efficient and robust techniques and technologies are required to handle such big data in terms of its acquisition, transmission, storage, computation, analytics and processing. This section discusses these aspects of big data processing.

5.1 Big Data Acquisition

As discussed in the previous sections, AV related data needs to be acquired through diverse sources, such as sensors, cameras, wireless communication technologies and in-vehicle network, in various V2X scenarios. It is important to ensure that the acquired data does not compromise information about driver or passengers. Therefore, ensuring security and privacy of the acquired data is crucial. The heterogeneity of data source presents several challenges in the data acquisition process, such as security and privacy issues, coverage area, cost to obtain data and corporate/government policies. Data scalability is a matter of concern in vehicular networks as the participants have limited resources. The computational infrastructure of the vehicle needs to be capable of storing and processing huge amounts of heterogeneous data [27, 28].

5.2 Big Data Preprocessing

AV data is acquired through heterogeneous sources at a high rate of data generation as seen in previous sections [28, 29]. Vehicular data obtained from different sources is in raw form, and needs to be processed before it can be brought to use. Data preprocessing can be carried out in five steps [30–32]:

1. **Data cleaning**
    a. Filling missing values
    b. Smoothening noisy data

   c. Confirmation and removal of outliers

2. **Data integration**

   a. Integrating data and converting into consistent data type

3. **Data transformation**

   a. Normalization, aggregation and summarization of data

4. **Data reduction**

   a. Reducing the size of the data

5. **Data discretization**

   a. Converting numerical values into categorical values

5.3 Big Data Storage

Traditional data storage systems such as Redundant Array of Independent Disks (RAID) are not suitable for vehicular networks and AVs because they are less reliable, less secure, and not designed to scale up to enormous data generated by these applications [33]. The data storage for AV and vehicular networks is required to have the following characteristics:

1. **Availability**: All time availability of data regardless of downtime
2. **Scalability**: Capable of storing the ever increasing amounts of data
3. **Security**: Data confidentiality and authenticated access over multiple drives, servers, containers and locations
4. **Efficiency:** Should be capable of managing Petabyte - Exabyte of storage
5. **Cost effectiveness**: Reduced cost of ownership and replication.

The following three types of storage is used for vehicular big data [34–36]:

1. **On-board storage**: Used to store data locally. Because of local storage, data can be accessed in real-time.
2. **Roadside storage**: Roadside infrastructure is preoccupied with storage. Data can be buffered and relayed for longer distances to the receiver that are not in direct communication range of the vehicle. Vehicle can also download data through roadside infrastructure. However, it adds to transmission delays. Location based advertisement of the nearby services use roadside infrastructure.
3. **Internet storage**: Vehicles communicating through internet to exchange important information such as weather, Over The Air (OTA) updates and infotainment, are some of the example use cases for this type of storage.

Based on access mechanisms, vehicular data storage can be classified into the following:

1. **Fast access**: Used for delay sensitive applications where data is accessed within a guaranteed delay. For example, data for vehicle safety and autonomous driving.
2. **Medium access**: Comprises of external storage that can be accessed through a reliable connection. For example, storage access through multi-hop communication and backhaul transmission. V2V assisted content delivery schemes, where the storage memory of a forwarding vehicle caches the popular content temporally until it comes across another forwarder is an example of medium access vehicular data storage.
3. **Slow access**: This is the external storage that can be accessed through an opportunistic connection due to mobility of vehicle and unavailability of consistent connection causing delays. Roadside infrastructure can be used for data storage for vehicles. Cloud storage is used to store data beyond local storage capacity and to improve utilization of overall vehicular resources. For example, V2R and V2C communication.

5.4 Big Data Analysis

As the idea of connected and autonomous vehicles may soon become a reality, researchers believe that the data analysis and storage requirements of AV pose challenge to the capabilities of most of the big data solutions available at present [37]. Analysis of the data produced and communicated for an AV needs to be analyzed and processed in real-time to safely navigate the vehicle. This helps in revealing hidden patterns, unknown correlations, driver behavior and preferences to make informed decisions in real-time [38]. The vehicle needs to choose from different data streams in real-time and analyze the most appropriate data stream based on the requirement. Therefore, an AV needs to run machine learning and analytics engines to recognize critical data and analyze it in real-time.

Hence, the data analytics and machine learning algorithms for AV should be able to:

1. Recognize critical data for a given scenario
2. Perform real-time data analysis
3. Compress and aggregate non-critical data for caching and future use
4. Periodically upload non-critical data to cloud for future analytics
5. Support download of required data from the cloud, process in real-time and act appropriately.

In levels 1–4 of autonomy supported in an AV, data analytics can guide the driver through driving decisions and safely navigate the vehicle depending on driver's behavior.

Such data analytics can be classified into three types:

1. **Descriptive analytics**: With the help of techniques such as data aggregation and data mining, descriptive analytics summarizes the historical data, which is used to identify behavioral patterns and relationships. This information provides vital

understanding of information about the behavior of different parameters across a variety of fields and industries.
In case of AV, descriptive analytics applies to moment-to-moment driving patterns and road behavior of the driver. This is aimed to study safe and risky behaviors of the driver.
2. **Predictive analytics**: While descriptive analytics identifies behavioral patterns based on historical data, predictive analytics is a method to identify the prospect of future outcomes. Predictive analytics assesses the future events by using data analytics and machine learning techniques on the historical data.
In case of AVs, based on the past driving patterns and road behavior, the vehicle can make informed judgement of driver's possible behavior in a given situation. For example, given the proximity with other vehicles or obstacles, whether the driver will slow down, overtake, navigate through or brake.
3. **Prescriptive analytics**: Based on the information obtained through the descriptive and predictive analytics, prescriptive analytics works towards establishing the actions to be taken in the future for a given scenario.

For the AV environment, prescriptive analytics uses the information obtained through descriptive and predictive analytics and issues recommendations based on driver's behavior in the past. The vehicle can assess the risk for a given situation and suggest appropriate action in real-time.

5.5 Big Data Computing

In an AV, most of the operational decisions need to be made by the vehicle instead of the driver (depending on the level of automation). In such a scenario, real-time decision making with reliability and security is paramount. To make real-time decisions, the level of functionality of automotive computing systems and computer processing needs to be increased [35, 37].

To enable AV functionality, the vehicles need to have the following characteristics:

1. **High computing power**: Approximately 1 GB of data will need to be processed and analyzed every second, and real-time decisions will need to be made within a fraction of second. This demands for high computing power at the vehicle.
2. **Centralized computing**: In most of the present day vehicles, different ECUs of the vehicle carry out their own computation, leading to a distributed computation architecture. As the decision making dependability among different ECUs increases, a centralized computing approach will be more efficient by reducing processing complexity, thus achieving real-time decisions.
3. **Small, high-processing units**: Multiple computing components will need to be installed at different places in the vehicles. Hence, these computing units will need to be small in size while providing high processing power.

4. **Security and privacy**: As applications communicate with roadside infrastructure, other vehicles and over cloud, it is essential to ensure that communicated data is secure and there are no malicious nodes in the communication link.

Edge Computing for Vehicular Networks

As discussed earlier, AVs generate and process huge volumes of data. Present mobile communication network architectures and cloud computing systems are not capable of handling such huge volumes of data. Network architecture based on topology aware computing and storage resources could be a solution for highly dynamic vehicular networks [38]. For such scenarios, topology-aware distributed clouds with multi operator edge computing capabilities is a solution suggested by the Automotive Edge Computing Consortium (AECC) [39]. Edge is the hierarchical distribution of non-central clouds and their computational resources in a flexible and topology aware manner. Real time applications require faster computing. In case a vehicle is not equipped with sufficient resources, it can use the resources available with the nearest regional edge infrastructure, instead of sending the computation request all the way to a central cloud. This edge infrastructure is distributed over the network at different locations/regions to cater to the needs of mobile nodes. Some architectures propose the use of computing resources of other vehicles (V2V) or of roadside infrastructure (V2R/V2I). Common services that can use edge computing for real-time data computing are intelligent/assisted driving, in-vehicle infotainment and OTA.

5.6 Big Data Transmission

As seen in previous sections, thousands of Giga Bytes of data is generated and exchanged per day by a single AV. In a network of such vehicles this data increases multiple times. Most of the vehicular data is real-time and needs technologies that support high bandwidth and data rate. Communication and networking technologies have to be capable and efficient enough to handle such huge data reliably and without delays (for real-time applications in particular). IEEE and 5GAA are the standardization bodies working in this direction. The protocols and standards used for vehicular communication, and their supported data rates are presented in Sect. 3. Density of vehicles in urban areas is high while their mobility is usually low (due to traffic congestion). While in rural areas, vehicle density is less and mobility is high. The study carried out in [40] presents the variation in bandwidth requirement of such different scenarios. As the number of AVs on roads increase in future, the demand for high bandwidth, data rates and efficient technologies will keep on increasing. Continuous efforts are being made by the research community to enhance efficiency and data rate of communication technologies.

Authors in [41] present the terahertz communication model for increase in network bandwidth. This model operates in 0.1–1 THz frequency band. Since this band is not used by many communication technologies, data transmission in this channel is relieved of interference. However, limitation of this channel is short range and line-of-sight communication, that makes this band unsuitable for vehicular communication. Vehicular networks use multi-hop communication. Hence, efficient cross-layer solutions (physical, Medium Access Control (MAC), and network layer solutions) are needed. Directional antenna/beamforming is used by the IEEE 802.11n, 802.11ac, 802.11ad (WiGig) to increase communication range and reduce interreference. However, using beamforming leads to the problem of deafness [42]. A cross-layer solution presenting the benefits of use of dual polarization with beamforming are presented in [43]. This solution simultaneously uses two orthogonal polarizations (vertical and horizontal polarizations) for data transmission. Since the data is transmitted over orthogonal polarizations, the signals do not interfere (as long as the orthogonality is maintained) while achieving data rates higher than cases where only single polarization is used. The Dual Polarized Directional Antenna (DPDA) based MAC layer solution is presented in [44]. The DPDA based multi-hop, multipath routing protocol is presented in [45].

Whether the application is connectionless (User Datagram Protocol (UDP) based) or connection oriented (Transmission Control Protocol (TCP) based), the traffic load on the network, affects the performance of the network [46]. Certain applications require higher data rates compared to others. For example, infotainment and navigation require higher data rates compared to location based commercial updates. Therefore, it is required to develop application-based transmission strategies. [47].

6 AV Security

An AV makes driving decisions based on its surroundings. It also communicates with other vehicles and roadside units in its vicinity. These vehicles have applications that require them to exchange information over edge/cloud computing platforms as well, as discussed in earlier sections. As more and more electronics, intelligence features and communication interfaces are integrated in the vehicle, the more vulnerable the vehicle becomes to malicious actions [48–50].

With increase in the number of ECUs in the vehicle, and availability of interfaces to access the ECU through wireless communication, in-vehicle network with access interfaces such as Universal Serial Bus (USB) and diagnostics port, it becomes easy to hack the vehicular components and networks. This also compromises the driver/passenger information of the vehicle [51].

The vehicle interfaces susceptible to exploits are as follows:

1. **Direct access interfaces**: Infotainment systems, Compact Disc (CD) player, on-board diagnostics
2. **Short-range wireless communication interfaces**: Remote keyless entry, tire pressure monitoring, Bluetooth, Wi-fi
3. **Long-range wireless communication interfaces**: Satellite radio, AM/FM radio, cellular communication, edge/cloud platform access.

A hacker can find a vulnerable ECU (which could be accessed remotely), reprogram it thus compromising the ECU, and send malicious content to other ECUs in the network through vehicle bus or other interfaces. This way the hacker can obtain control on the vehicle. Table 3 shows the list of threats on vehicular networks.

Some of the techniques that can be used to mitigate these attacks over V2X and IVN systems are:

1. Message encryption and authentication
2. Firewall to restrict communication between networks
3. Securing the gateway modules that are used to facilitate communication between different IVN protocol networks
4. Intrusion detection and prevention systems to detect and prevent malicious content
5. Securing the safety critical systems such as braking, engine control unit and steering control unit

Table 3 Threats on vehicular networks

Threats to vehicle	Method	Possible interfaces to access the network
Man in the middle (MITM) attack	Intercept, modify and resend the information	CAN/BT/WiFi/Cellular/OBD
Denial of service (DoS) attack	Delete encrypted premium content Flood the network	Unauthorized applications from remote device or downloaded to the infotainment system
Replay attack	Replay V2X messages Replay in-vehicle messages	V2X/IVN/OBD
Collect private information	Record vehicle messages Track vehicle's location and Transactions Infer private information about driver and passengers	V2X/GPS
Unauthorized control of vehicle parameters	Install program onto vehicle's IVN bus	OBD/USB
False alerts	Transmission of false hazard warnings Injecting RDS-TMC (radio data system-traffic message channel) traffic information signals	RDS/TMC/V2V
Conceal location information	Use GPS jamming device preventing fleet owner from tracking the fleet	GPS
Tamper ECU data	Change ECU configuration Modify calibration file Tamper persistent database	OBD/CAN/JTAG
Bluejacking and Bluebugging	Send unsolicited messages over Bluetooth	BT

6. Securing the infotainment systems such as telematics, in-vehicle infotainment (IVI) and radio system
7. Secure firmware update
8. Secure hardware.

7 Research Avenues

Vehicular networks and AVs are the upcoming technologies and have caught interest of numerous researchers all over the world. The research on big data in AV primarily deals with transmission of big data, studying the available data, storage and computation, and data security. This section presents some of the research avenues and related research works in this field.

7.1 *Big Data Transmission*

The communication technologies for AV need to support high data rate and bandwidth to allow real-time communication in different V2X scenarios. Research community is constantly working towards developing solutions based on standard vehicular communication standards. Physical, MAC and network layer routing solutions are needed for efficient bandwidth usage and interference avoidance, channel access and routing of information over a network respectively. Many of the existing solutions have been discussed in Sects. 3 and 5. Some of the latest research avenues are discussed here:

Information Centric Networking (ICN): Traditional networking communication is address based, wherein the source (client) sends request to the destination (server) and receives a reply. In highly dynamic vehicular networks, information is given importance, instead of where it is coming from. Information such as real-time vehicle status, road conditions, weather information (usually available in the cloud), is required with minimal delay. ICN allows intermediate nodes or roadside infrastructure to store this information. On receiving the query for such information, a node or roadside infrastructure having latest information for the query replies with information, instead of sending the query all the way to server/cloud [52, 53]. This drastically reduces delay and helps in efficient use of bandwidth. Content Centric Networking (CCN) is also referred to as Named Data Networking (NDN) and is one form of ICN that supports name-based data retrieval and pervasive caching. The concept of NDN applied to vehicular networks has given rise to Vehicular-NDN (V-NDN) or Named Data Vehicular Networks (NDVN) [54, 55]. In this field, content naming, caching and forwarding schemes need to be developed.

Some of the open research issues in this domain are:

1. **Development of routing strategies**: Efficient routing mechanisms needed to fulfil QoS requirements of vehicular network.

2. **Content naming schemes**: Appropriate content naming schemes need to be developed for vehicular networks.
3. **Caching strategies**: Intermediate nodes and roadside units can cache the data, but there are no standard schemes about storing and forwarding of data.
4. **Interest and data flooding**: Method and periodicity for flooding of interest and data packets needs to be standardized.
5. **Approach to handle dynamic network topologies**: Topologies in vehicular networks change constantly leading to disruption of connection. Solutions to overcome this problem are required.
6. **Security in V-NDN/NDVN**: Authentication and encryption add substantial overhead. There is a need to develop light weight solutions for secure communication [56].

Software Defined Networking (SDN) and Network Function Virtualization (NFV)

Big data generated by AVs and vehicular networks will need high bandwidth, storage and computing power. The individual vehicle may lack enough storage and computing power for all the generated data. In such cases, edge and cloud computing, and virtualization of network resources will play a crucial role in reducing CAPEX and OPEX of vehicular networks [57–59]. Traditional network technology is not efficient enough to handle the volumes of data generated by AVs. SDN allows dynamic configuration of network resources based on the need and achieves better QoS in a vehicular network scenario when compared to the traditional networking architecture. Many researchers have proposed Software Defined Vehicular Network (SDVN) architectures [60]. Some of the benefits of SDVNs are dynamic network configuration, better resource utilization, reduced latency and integration of heterogeneous networks through network controller.

While SDVNs provide these benefits, the highly dynamic nature of vehicular networks along with huge data produced and communicated over wireless medium leads to the following challenges, with need to develop solutions:

1. **Mobility management**: This leads to varying channel conditions. SDN paradigm needs to be modified to handle the issues of network mobility and connection breakages. To ensure constant network connectivity and maintain required QoS, the SDN based solutions need to predict the driving patterns and pre-allocate resources while handling the concerns of privacy.
2. **Lack of standardized APIs to handle heterogeneous network traffic**: SDN network controller provides separation between applications and network infrastructure. However, standardized eastbound/westbound APIs and northbound APIs need to be developed for different data generating applications for successful integration of network traffic.
3. **Vulnerability of SDN controller**: In any SDN based solution, a compromised network controller can prove disastrous for the network leading to Distributed Denial of Service (DDoS) attack.

4. **Security**: The presence of malicious vehicles and RSUs can lead to routing-based attacks such as blackhole, sinkhole, sybil and replay attacks. Malware attack injection can lead to replication of malicious software on network controllers, routers, switches, vehicles and RSUs leading to tampering of forwarding rules for resource allocation.

7.2 Machine Learning (ML) for Vehicular Networks

Presence of big datasets, need for high computing capacity, constantly changing network topologies and channel conditions, and need to serve real-time applications in vehicular networks demands for intelligent learning algorithms that can predict data and network conditions with accuracy and make faster decisions. ML helps to model complex network environments, obtain abstract features, and make appropriate decisions to achieve required QoS for AV [61–63].

In the vehicular networks, traffic prediction in ML algorithms can help with the following:

1. **Network queuing analysis**: Large scale network queuing analysis helps in uniform understanding of 'congestion propagation' among participating nodes and achieve an optimal solution to relieve congestion. There is a need to develop solutions to define congestion threshold, detect queue evolution pattern, congestion control and routing to avoid/overcome congestion.
2. **Analysis of big data transmission dynamics**: This is used to develop solutions to predict routing delays due to big data flows, average link quality and stability, load balancing over multiple paths and nodes to avoid bottleneck.
3. **ML for SDVN**: The network controller in SDVN can make decisions about resource allocation and configure the network based on the network traffic, congestion and delays. ML can be used to help the network controller make these decisions by studying the traffic patterns. ML integrated with SDVN solutions helps in achieving better network prediction and take timely action.

7.3 Vehicular Network Security

Any network and communication is vulnerable to attacks and needs security. Vehicular networks more so, because of primary use of wireless medium for communication and mobility of vehicular nodes. In a VANET, nodes join and leave the network on the go and carry out V2X communication. Remote communication combined with access to vehicles and their units makes the vehicles more vulnerable to security attacks [64–66]. In Sect. 6, different threats to vehicular networks and possible attack surfaces are presented. This section discusses the open research issues in the domain of vehicular network security.

Table 4 Open security issues

	Issues			
Levels	Prevent access	Detect attacks	Reduce impact	Fix vulnerabilities
Secure processing	Authentication code, secure boot	Run-time integrity protection	Resource control, virtualization	
Secure network	Secure messaging			Secure OTA updates
Secure gateway	Firewalls, context aware message filtering	Intrusion detection systems (IDS)	Separate functional domains, isolated transmission control unit (TCU) and on-board diagnostics (OBD)	Secure firmware
Secure interfaces	Machine-to-machine (M2M) authentication Isolation of access points			

Open security issues at different levels of AV: AV security can be broadly segregated into security of processing units, network, gateway and interfaces. Table 4 presents the required possible solutions for open security issues at these levels.

Secure processing: The ECUs on in-vehicle network and the gateway nodes need to process messages exchanged among them using different communication protocols. The messages could belong to the critical safety domain as well as not so critical comfort domain. However, one malignant message can corrupt the whole network. To overcome this, solutions are required for secure vehicle data processing that could comprise of secure boot, ECU level message encryption and authentication, key storage and secure OTA software updates. Secure firmware updates with minimal vehicle downtime during update or failed updates may leave the vehicle unusable.

Secure network: Some of the open research issues to attain network communication in AV are context-aware message filtering, message protection through encryption and authentication, network component (ECU, OBU, RSU, gateway) authentication and secure end-to-end connection establishment. The wireless network used for V2X communication too are vulnerable to several attacks as discussed in Sect. 6. Solutions to avoid or overcome these attacks in the highly dynamic vehicular environment is essential, but difficult to achieve.

Secure gateway: Gateway is an essential component of the vehicular network that enables communication between heterogeneous networks using different communication protocols. It provides network isolation and security between functional domains and networks. A compromised gateway module can corrupt the whole network. Hence, a secure gateway is essential. Open research issues for secure network are as important to achieve a secure gateway. A message filter or a firewall solution at the gateway module can isolate the network from malicious data.

Gateway should be capable of key storage along with encryption and authentication of messages.

Secure interfaces: An interface is the entry/exit point of the data to/from the network. Components that are used for M2M and V2X communication, telematics, infotainment and diagnostics, if not secured, can leave the entry of malicious data in the network undetected. Hence, it is very important to secure these interfaces. Secure interfaces need to be implemented on a tamper-resistant platform so that they can securely host security applications and their confidential data, thus staying protected against physical attacks. Solutions to achieve secure interfaces need to aim for secure crypto processing, crypto key generation and storage, and secure certificate handling to validate and store information.

8 Conclusion

This chapter presents the significance, applications, related technologies, issues and research avenues in the domain of big data for AVs. The chapter starts with an introduction to vehicular communication (Intra-vehicle and Inter-vehicle communication). Protocols and standards for V2X communication are presented and explained. AV use cases are discussed along with enabling protocols, components and technologies. The big data sources in vehicular networks, rate of data generation, communication and processing are tabulated and discussed, thus highlighting the significance of big data technology in vehicular networks. The chapter presents ongoing research with respect to the technologies involved in big data processing in AV such as data transmission, acquisition, storage, computing, analytics and processing. The highly dynamic nature of vehicular networks presents a major concern for security and privacy on big data in AV. This chapter discusses these security issues and possible solutions. Big data in AV is an emerging field having diverse research avenues. Research directions and open research issues in big data transmission, ML for vehicular networks and vehicular network security are presented with an aim to help the researchers interested in pursuing research in this field.

References

1. Rakouth H, Alexander P, Kosiak W, Fukushima M, Ghosh L, Hedges C, Kong H, Kopetzki S, Siripurapu R, Shen J (2013) V2X communication technology: field experience and comparative analysis. In: Proceedings of the FISITA 2012 world automotive congress. Springer, Berlin, Heidelberg, pp 113–129
2. Sładkowski A, Pamuła W (eds) (2016) Intelligent transportation systems-problems and perspectives, vol 303. Springer International Publishing
3. Mangel T, Klemp O, Hartenstein H (2011) 5.9 GHz inter-vehicle communication at intersections: a validated non-line-of-sight path-loss and fading model. EURASIP J Wirel Commun Netw 2011(1), 182

4. Zeadally S, Hunt R, Chen YS, Irwin A, Hassan A (2012) Vehicular ad hoc networks (VANETS): status, results, and challenges. Telecommun Syst 50(4):217–241
5. Mahmud SM, Alles S (2005) In-vehicle network architecture for the next-generation vehicles. SAE Trans:466–475
6. Khaled Y, Ducourthial B, Shawky M (2005, May) IEEE 802.11 performances for inter-vehicle communication networks. In: 2005 IEEE 61st vehicular technology conference, vol 5. IEEE, pp 2925–2929
7. V2V safety communication, Available https://www.sae.org/standards/
8. 5G automotive association, Available https://5gaa.org/
9. MacHardy Z, Khan A, Obana K, Iwashina S (2018) V2X access technologies: regulation, research, and remaining challenges. IEEE Commun Surv Tutor 20(3):1858–1877
10. Mohammad SA, Rasheed A, Qayyum A (2011, Mar) VANET architectures and protocol stacks: a survey. In: International workshop on communication technologies for vehicles. Springer, Berlin, Heidelberg, pp 95–105
11. Kenney JB (2011) Dedicated short-range communications (DSRC) standards in the United States. Proc IEEE 99(7):1162–1182
12. Li YJ (2010, Nov) An overview of the DSRC/WAVE technology. In: International conference on heterogeneous networking for quality, reliability, security and robustness. Springer, Berlin, Heidelberg, pp 544–558
13. Gallardo JR, Makrakis D, Mouftah HT (2010) Mathematical analysis of EDCA's performance on the control channel of an IEEE 802.11 p WAVE vehicular network. EURASIP J Wirel Commun Netw 2010(1):489527
14. van Eenennaam M, van de Venis A, Karagiannis G (2012, Nov) Impact of IEEE 1609.4 channel switching on the IEEE 802.11 p beaconing performance. In: 2012 IFIP wireless days. IEEE, pp 1–8
15. Weil T (2009, Nov) Service management for ITS using WAVE (1609.3) networking. In: 2009 IEEE Globecom workshops. IEEE, pp 1–6
16. Rabadi NM (2010, Nov) Implicit certificates support in IEEE 1609 security services for wireless access in vehicular environment (WAVE). In: The 7th IEEE international conference on mobile ad-hoc and sensor systems (IEEE MASS 2010). IEEE, pp 531–537
17. Park H, Miloslavov A, Lee J, Veeraraghavan M, Park B, Smith BL (2011) Integrated traffic–communication simulation evaluation environment for intellidrive applications using SAE J2735 message sets. Transp Res Rec 2243(1):117–126
18. Rostami A, Krishnan H, Gruteser M (2018, June) V2V safety communication scalability based on the SAE J2945/1 standard. In: Proceedings of ITS American Annual Meeting
19. Qu Q, Li B, Yang M, Yan Z, Yang A, Deng DJ, Chen KC (2019) Survey and performance evaluation of the upcoming next generation WLANS standard-IEEE 802.11 ax. Mobile Netw Appl 24(5):1461–1474
20. Papathanassiou A, Khoryaev A (2017) Cellular V2X as the essential enabler of superior global connected transportation services. IEEE 5G Tech Focus 1(2):1–2
21. Friesen MR, McLeod RD (2015) Bluetooth in intelligent transportation systems: a survey. Int J Intell Transp Syst Res 13(3):143–153
22. Prakash P, Sharma R, Sindhu S, Shankar T (2017, Dec) Visible light communication using solar panel. In: 2017 2nd international conference on emerging computation and information technologies (ICECIT). IEEE, pp 1–5
23. Winning A (2019) Number of automotive ECUs continues to rise. Available https://www.eenewsautomotive.com/news/number-automotive-ecus-continues-rise
24. https://www.synopsys.com/automotive/autonomous-driving-levels.html
25. Daniel A, Subburathinam K, Paul A, Rajkumar N, Rho S (2017) Big autonomous vehicular data classifications: towards procuring intelligence in ITS. Vehic Commun 9:306–312
26. Nelson P (2016) Just one autonomous car will use 4,000 GB of data/day. Available https://www.networkworld.com/article/3147892/one-autonomous-car-will-use-4000-gb-of-dataday.html
27. Lyko K, Nitzschke M, Ngomo ACN (2016) Big data acquisition. In: New horizons for a data-driven economy. Springer, Cham, pp 39–61

28. Brinkmann BH, Bower MR, Stengel KA, Worrell GA, Stead M (2009) Large-scale electrophysiology: acquisition, compression, encryption, and storage of big data. J Neurosci Methods 180(1):185–192
29. Cho W, Choi E (2017) Big data pre-processing methods with vehicle driving data using MapReduce techniques. J Supercomput 73(7):3179–3195
30. Fiosina J, Fiosins M, Müller JP (2013) Big data processing and mining for next generation intelligent transportation systems. J Teknologi 63(3)
31. Pandey MK, Subbiah K (2016, Dec) Social networking and big data analytics assisted reliable recommendation system model for internet of vehicles. In: International conference on internet of vehicles. Springer, Cham, pp 149–163
32. Guerreiro G, Figueiras P, Silva R, Costa R, Jardim-Goncalves R (2016, Sept) An architecture for big data processing on intelligent transportation systems. An application scenario on highway traffic flows. In: 2016 IEEE 8th international conference on intelligent systems (IS). IEEE, pp 65–72
33. Yang F, Li J, Lei T, Wang S (2017) Architecture and key technologies for internet of vehicles: a survey. J Commun Inf Netw 2(2):1–17
34. Hu Z, Zheng Z, Wang T, Song L, Li X (2017) Roadside unit caching: auction-based storage allocation for multiple content providers. IEEE Trans Wireless Commun 16(10):6321–6334
35. Zhang W, Zhang Z, Chao HC (2017) Cooperative fog computing for dealing with big data in the internet of vehicles: architecture and hierarchical resource management. IEEE Commun Mag 55(12):60–67
36. Mekki T, Jabri I, Rachedi A, ben Jemaa M (2017) Vehicular cloud networks: challenges, architectures, and future directions. Vehic Commun 9:268–280
37. Darwish TS, Bakar KA (2018) Fog based intelligent transportation big data analytics in the internet of vehicles environment: motivations, architecture, challenges, and critical issues. IEEE Access 6:15679–15701
38. Al Najada H, Mahgoub I (2016, Dec) Anticipation and alert system of congestion and accidents in VANET using big data analysis for intelligent transportation systems. In: 2016 IEEE symposium series on computational intelligence (SSCI). IEEE, pp 1–8
39. Automotive Edge Computing Consortium (AECC) (2020) General principle and vision, white paper. Available https://aecc.org/wp-content/uploads/2020/03/AECC_White_Paper_v3.0_-_Jan_31_2020.pdf
40. Thriveni HB, Kumar GM, Sharma R (2013, Apr) Performance evaluation of routing protocols in mobile ad-hoc networks with varying node density and node mobility. In: 2013 international conference on communication systems and network technologies. IEEE, pp 252–256
41. Zhang C, Ota K, Jia J, Dong M (2018) Breaking the blockage for big data transmission: gigabit road communication in autonomous vehicles. IEEE Commun Mag 56(6):152–157
42. Sharma R, Kadambi G, Vershinin YA, Mukundan KN (2016) A survey of MAC layer protocols to avoid deafness in wireless networks using directional antenna. In: Mobile computing and wireless networks: concepts, methodologies, tools, and applications. IGI Global, pp 1758–1797
43. Rinki S (2014) Simulation studies on effects of dual polarisation and directivity of antennas on the performance of MANETs. In: Doctoral dissertation, Ph. D. thesis, Coventry University, UK
44. Sharma R, Kadambi GR, Vershinin YA, Mukundan KN (2015, Apr) Dual polarised directional communication based medium access control protocol for performance enhancement of MANETs. In: 2015 Fifth international conference on communication systems and network technologies. IEEE, pp 185–189
45. Sharma R, Kadambi GR, Vershinin YA, Mukundan KN (2015, Apr) Multipath routing protocol to support dual polarised directional communication for performance enhancement of MANETs. In: 2015 fifth international conference on communication systems and network technologies. IEEE, pp 258–262
46. Gopinath T, Kumar AR, Sharma R (2013, Apr) Performance evaluation of TCP and UDP over wireless ad-hoc networks with varying traffic loads. In: 2013 international conference on communication systems and network technologies. IEEE, pp 281–285

47. Huang Y, McMurran R, Amor-Segan M, Dhadyalla G, Jones RP, Bennett P, Mouzakitis A, Kieloch J (2010) Development of an automated testing system for vehicle infotainment system. Int J Adv Manuf Technol 51(1–4):233–246
48. Xue M, Wang W, Roy S (2014) Security concepts for the dynamics of autonomous vehicle networks. Automatica 50(3):852–857
49. Amoozadeh M, Raghuramu A, Chuah CN, Ghosal D, Zhang HM, Rowe J, Levitt K (2015) Security vulnerabilities of connected vehicle streams and their impact on cooperative driving. IEEE Commun Mag 53(6):126–132
50. Thing VL, Wu J (2016, Dec) Autonomous vehicle security: a taxonomy of attacks and defences. In: 2016 IEEE international conference on internet of things (iThings) and IEEE green computing and communications (GreenCom) and IEEE cyber, physical and social computing (CPSCom) and IEEE smart data (SmartData). IEEE, pp 164–170
51. Yu L, Deng J, Brooks RR, Yun SB (2015, Apr) Automobile ecu design to avoid data tampering. In: Proceedings of the 10th annual cyber and information security research conference, pp 1–4
52. Mau DO, Zhang Y, Taleb T, Chen M (2014, May) Vehicular inter-networking via named data-an opnet simulation study. In: International conference on testbeds and research infrastructures. Springer, Cham, , pp 116–125
53. Khelifi H, Luo S, Nour B, Moungla H, Faheem Y, Hussain R, Ksentini A (2019) Named data networking in vehicular ad hoc networks: state-of-the-art and challenges. In: IEEE Commun Surv Tutor
54. TalebiFard P, Leung VC, Amadeo M, Campolo C, Molinaro A (2015) Information-centric networking for VANETs. In: Vehicular ad hoc networks. Springer, Cham, pp 503–524
55. Xiao J, Deng J, Cao H, Wu W (2016, Dec) Road segment information based named data networking for vehicular environments. In: International conference on algorithms and architectures for parallel processing. Springer, Cham, pp 245–259
56. Jain V, Kushwah RS, Tomar RS (2019) Named data network using trust function for securing vehicular ad hoc network. In: Soft computing: theories and applications. Springer, Singapore, pp 463–471
57. Ananth MD, Sharma R (2016, Dec) Cloud Management using network function virtualization to reduce CAPEX and OPEX. In: 2016 8th international conference on computational intelligence and communication networks (CICN). IEEE, pp 43–47
58. Ananth MD, Sharma R (2017, Jan) Cost and performance analysis of network function virtualization based cloud systems. In: 2017 IEEE 7th international advance computing conference (IACC). IEEE, pp 70–74
59. Sharma R, Reddy H (2019, Dec) Effect of load balancer on software-defined networking (SDN) based cloud. In: 2019 IEEE 16th India council international conference (INDICON). IEEE, pp 1–4
60. Alioua A, Senouci SM, Moussaoui S (2017, June) dSDiVN: a distributed software-defined networking architecture for infrastructure-less vehicular networks. In: International conference on innovations for community services. Springer, Cham, pp 56–67
61. Taherkhani N, Pierre S (2016) Centralized and localized data congestion control strategy for vehicular ad hoc networks using a machine learning clustering algorithm. IEEE Trans Intell Transp Syst 17(11):3275–3285
62. Grover J, Prajapati NK, Laxmi V, Gaur MS (2011, July) Machine learning approach for multiple misbehavior detection in VANET. In: International conference on advances in computing and communications. Springer, Berlin, Heidelberg, pp 644–653
63. Kulkarni SA, Rao GR (2010, Aug) Vehicular ad hoc network mobility models applied for reinforcement learning routing algorithm. In: International conference on contemporary computing. Springer, Berlin, Heidelberg, pp 230–240
64. Xu L, Yu X, Wang H, Dong X, Liu Y, Lin W, Wang X, Wang J (2019) Physical layer security performance of mobile vehicular networks. In: Mobile networks and applications, pp 1–7
65. Moustafa H, Bourdon G, Gourhant Y (2006, May) Providing authentication and access control in vehicular network environment. In: IFIP international information security conference. Springer, Boston, MA, pp 62–73

66. Franeková M, Lüley P (2013, Oct) Security of digital signature schemes for car-to-car communications within intelligent transportation systems. In: International conference on transport systems telematics. Springer, Berlin, Heidelberg, pp 258–267

Deep Learning and Object Detection for Safe Driving

Analysis of Target Detection and Tracking for Intelligent Vision System

K. Kalirajan, K. Balaji, D. Venugopal, and V. Seethalakshmi

Abstract In the present scenario, most of the researchers are highly motivated to do new findings in the field of computer vision because of its tremendous applications such as video surveillance, human-machine interaction, traffic monitoring, human behavioral analysis, the guided missile, and military services, medical applications, and vehicle navigation. A typical video data frame comprises both foregrounds as well as background information. The pixel points that describe the target features in the region of interest are considered foreground information, and the rest of the feature points are treated as background information. Moving object detection plays a pivotal role in any kind of computer vision applications. Moving object detection is the process of identifying the class objects such as people, vehicles, toys, and human faces in the video sequences more precisely without background disturbances. In most cases, the existing moving object detection approaches concentrate only on the foreground information and frequently ignore the background information. As a result, trackers will be deviated away from the target and detect the non-foreground objects. Recently, several contributions have been proposed for moving object detection. However, the robustness and novelty are still challenging to achieve because of the complex environments, including illumination changes, rapid variations in target appearance, similar objects in the background, occlusions, target rotations, scaling, fast and abrupt motion changes, moving soft shadow, flat surface regions, and dynamic backgrounds. This book chapter introduces the concept of an intelligent video surveillance system and the major challenges involved in moving object detection. This chapter also deliberates the related backgrounds and their shortcomings in different perspectives. It also presents potential algorithms for efficient object detection and tracking. Finally, the book chapter is concluded with research opportunities pertaining to object detection and tracking.

K. Kalirajan (✉) · D. Venugopal · V. Seethalakshmi
KPR Institute of Engineering and Technology, Arasur, Coimbatore, Tamil Nadu, India
e-mail: ktvr.rajan@gmail.com

K. Balaji
SNS College of Engineering, Coimbatore, Tamil Nadu, India

K. R. Ahmed et al. (eds.), *Deep Learning and Big Data for Intelligent Transportation*, Studies in Computational Intelligence 945,
https://doi.org/10.1007/978-3-030-65661-4_3

Keywords Object detection and tracking · Video surveillance · Feature extraction · Video analytics

1 Introduction

Intelligent transportation is the need for the hour in the present scenario to effectively handle traffic congestion and transportation planning. In real-world traffic monitoring, detection, and tracking of vehicles in monitoring scenes are required in all—weather conditions at all times. This kind of vehicle tracking can automatically detect new targets or the disappearance of existing targets. Video surveillance is an emerging field with more advent for anonymous activity monitoring in the restricted areas, and it significantly becomes a part of life today. Usually, the video surveillance systems observe and analyze the massive amount of visual information to determine the suspicious activities in the given image frame. However, it is difficult to store and analyze the vital surveillance data manually due to boredom and exhaustion. Alternatively, an intelligent video surveillance system can support manual operations in event detection and other activity-based analysis. Figure 1 shows the basic building blocks of an intelligent vehicle tracking system.

Moving object detection is the first and foremost step in any kind of video processing applications such as video surveillance, traffic monitoring, human-computer interaction, people monitoring, military and border security service, and vehicle navigations etc. The moving object detection is the process of identifying and locating the target more appropriately in all image frames without background intervention. Any particular region of interest within a frame such as a vehicle, animal, people and other moving objects can be considered as a target to be identified by the vision system. Commonly used moving object detection methods include background subtraction, a mixture of Gaussians, an adaptive mixture of Gaussians, optical flow based approach, and temporal differencing. Since the performance of moving object detection algorithm is greatly influencing the high-level abstractions such as classification, tracking, and event analysis, an intelligent video surveillance system requires more appropriate and robust object detection algorithms. Object classification is the second step in smart video surveillance system. It will classify the objects into people, vehicle, animal and other targets such as toys, buildings etc. Object classification will further make the subsequent processes as more reliable. Normally, the object classification is based on the spatial features and temporal features. Sometimes both spatial and temporal features are used for better performance. Object tracking is the process of creating the temporal connectivity between the previous and current

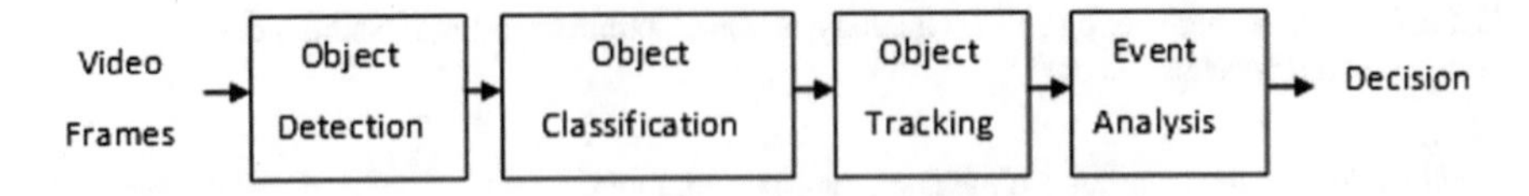

Fig. 1 Typical intelligent vehicle tracking system

object frames. Usually, it keeps the track of the detected objects in the successive frames by tracking their trajectories, motions, and orientations. It is necessary to assist the higher level processes in the video surveillance system. In event analysis, the actions and behaviors of the tracked objects will be analyzed and described in the final decision. The optimal and feasible decision about the event will lead to early effective precautions and corrective measures to avoid the event before occurring. This kind of elegant video surveillance system finds the applications in various fields as mentioned below.

- Public security in banking, supermarket, and railway stations etc.
- Old age home monitoring, and remote medical assistance
- Traffic monitoring, vehicle navigation and intelligent car parking system etc.
- Abandoned object detection, Guided missiles, border security, and terrorist activity monitoring etc.

1.1 Major Challenges Involved in Object Detection and Tracking

A smart video surveillance system demands robust moving object detection and tracking for high-level video analysis. In particular, moving object detection is mainly intended to discriminate the foreground objects from the background information with complete rid of background influences. However, the complex and dynamic background conditions make the object detection process very tricky. Major challenges involved in moving object detection process (see Fig. 2) are listed below.

- Illumination changes and challenging weather conditions
- Dynamic background
- Partial and/or full occlusion
- Background Clutter and similar object regions
- Moving cast shadow
- Abrupt motion changes, rotations, and Scale changes
- Image noises due to aperture errors and camera jitter etc.

Illumination Changes and Challenging Weather Conditions

An illumination change in background environments strongly affects the performance of object detection and tracking system. In outdoor environments, it occurs due to sunlight variations whereas, in indoor environments, the artificial lighting conditions are the root cause. Additionally, the tricky weather conditions such as mist and cloudy can cause the illumination changes. An example of illumination change is shown in Fig. 2a.

Fig. 2 Example for major challenges involved in moving object detection: **a** Illumination changes, **b** object occlusion, **c** abrupt motion changes, **d–f** in-plane object rotations, **g–h** scale changes, **i** dynamic background, **j** moving cast shadow and **k** out-of-plane rotations. *Source* http://cvlab.hanyang.ac.kr/tracker_bench [1], http://www4.comp.polyu.edu.hk/~cslzhang/FCT/ FCT. html [2], http://perception.i2r.a-star.edu.sg/bk_model/bk_index.html [3]

Partial and/or Full Occlusion

Occlusion is another challenging condition in object detection and tracking. The foreground objects are partially or completely occluded by other non-target objects during object detection and tracking (see Fig. 2b). Under such conditions, no information is available for target modeling which results in non-target object detection. As a result, it is an essential to building up an effective algorithm to detect the target when it reappears.

Background Clutter

The cluttered background is another important factor which degrades the system performance. Due to background clutter, it is more difficult to discriminate foreground objects from the background.

Abrupt Motion Changes, Rotations, and Scale Changes

Abrupt motion changes, rotations and scale changes in target observation are the significant challenging parameters in object detection and tracking. Due to the abrupt motion changes such as a jump, run, pull or a push, the target will undergo severe appearance changes and motion blurring (see Fig. 2c). The fast moving object creates a trail of ghost region and the irregular motion produces the artifacts. Identification and localization of such target will become complex. Furthermore, the rotations (see Fig. 2d–f, k) and scale changes (see Fig. 2g–h) significantly affect the pixel relationships and their orientations. The traditional methods cannot hope with these complex situations.

Similar Object Region and Dynamic Background

Moving object detection in presence of foreground objects with similar background color is also a challenging job (See Fig. 2g–h, j). Robust object detection algorithm must be able to deal with these challenges. In dynamic background videos, the background details such as rippling water, waving trees, waving clouds in the sky and waving curtains in the room are the time varying quantities which may neither regular nor irregular and they often lead to vulnerable target detection. Figure 2i shows an example of dynamic background image frame.

Moving Cast Shadow

Shadow removal is one among the major issues in object detection. Basically, a shadow is created because of changing the lighting conditions. Depending on the viewpoint, the shadow can be categorized into static shadow and dynamic shadow. Static shadow occurs due to static objects whereas dynamic shadow is formed by the moving objects. In real world scenes, the moving cast shadows will distract the target tracker which results in a false detection. For example, in traffic monitoring, the moving cast shadow appears as a part of the vehicle and the bounding box detects this shadow as a car (See Fig. 2j). The usage of illumination invariant features can increase the system efficacy.

Image Noises

Video frames are sometimes corrupted by the random noises such as Gaussian noise, shot noise, and salt-pepper noise. The image quality may also be affected by the aperture errors in the camera sensor, camera jitter, and low illumination etc. Since picture quality directly influences the performance of visual analysis, the object detection and tracking algorithms must be able to handle these noises explicitly.

2 Moving Object Detection Methods

There are several methods to detect the moving object regions in successive video frames. The following sections describe the various methods of moving object detection.

2.1 Temporal Difference Method

In static environments, the background is fairly stable and it is easy to estimate the object motion by subtracting the current frame with the stationary background model. On the other hand, in the dynamic environment, the background is continuously changing with moving camera and the initial background model is not enough to estimate the moving object motion. The temporal difference method considers two or three consecutive frames to estimate the object motion. Since the recent image frames are used for motion estimation, this method is more appropriate for dynamic environments. Types of temporal difference method includes inter frame difference technique and three frame difference technique. In inter frame difference technique detects the target by differencing the two successive frames. Let $g(x, y)$ be an initial video frame and $g(x, y, t + 1)$ be the next frame. Then, the moving object is calculated as follows:

$$D(x, y) = g(x, y, t + 1) - g(x, y) \tag{1}$$

$$ROI = \begin{cases} 1;\ D(x, y) > T_h \\ 0;\ D(x, y) \le T_h \end{cases} \tag{2}$$

where $D(x, y)$, is the difference image, T_h is the threshold for object classification, and ROI is the foreground region. From Eq. (2), it can be seen that the feature points that are lower than the threshold value, are considered as background and all other feature points are treated as a foreground object. The inter-frame differencing methods are strong enough to produce well-known motion blob in the difference image for fast moving objects. However, the motion blobs of noise and foreground object are not distinguishable for slowly moving objects. Moreover, the two frame differencing is greatly affected by the shadowing problem. In three frame difference method, three consecutive images are used instead of two image frames to calculate the difference image. Let $g_1(x, y)$, $g_2(x, y)$ and $g_3(x, y)$ be three consecutive image frames. The difference image $D(x, y)$ is calculated as follows.

$$D(x, y) = \min\{|(g_1(x, y) - g_2(x, y))|, |(g_2(x, y) - g_3(x, y))|\} \tag{3}$$

Then, the thresholding is done on the difference image as similar to Eq. (2) to detect the foreground objects. The three frame differencing method is best suited for the dynamic environments. However, it suffers from the object outlier problem. Furthermore, it is very sensitive to the threshold value and the target completely vanishes in the case of a larger threshold value. Additionally, it fails to detect the targets that are stop moving temporarily.

2.2 Simple Background Subtraction

In simple background subtraction method, the moving object is detected by subtracting the current image frame from the background model. Let $f_t(x, y)$ and $Bg_t(x, y)$ be the current image frame, and the background model respectively. The difference image $D_t(x, y)$ is calculated according to the Eq. (4).

$$D_t(x, y) = f_t(x, y) - Bg_t(x, y) \tag{4}$$

The moving object $BI_t(x, y)$ is calculated by the decision threshold T_h as follows [4]:

$$BI_t(x, y) = \begin{cases} 0; \ D_t(x, y) > T_h \\ 1; \ D(x, y) \leq T_h \end{cases} \tag{5}$$

In background subtraction with a dynamic threshold, the threshold value is dynamically updated according to the external environment conditions. Let ΔT be the dynamic threshold, then the Eq. (5) can be rewritten as follows:

$$BI_t(x, y) = \begin{cases} 0; \ D_t(x, y) > T_h + \Delta T \\ 1; \ D(x, y) \leq T_h + \Delta T \end{cases} \tag{6}$$

where ΔT is given by

$$\Delta T = \alpha \frac{1}{M \times N} \sum_{x=0}^{N-1} \sum_{y=0}^{M-1} f_t(x, y) - Bg_t(x, y) \tag{7}$$

Here, α is the scaling factor and $(M \times N)$ is the size of the image frame. Background subtraction methods are most widely used for moving object detection with fixed camera motion. Zhang and Liang [5] updated a consistent background model and identifies the moving object by a dynamic threshold. Morphological operations are used to remove the background influences. The background subtraction methods can provide the complete target information for well-known background model. However, it is sensitive to dynamic background variations and camera motion changes.

2.3 Background Modeling Approach

In background modeling, the desirable characteristics of each pixel are learned and updated in all image frames and thereby improves the results of background subtraction under dynamic environments. Single Gaussian model and mixture-of-Gaussian model are the most popular approaches for background modeling.

2.3.1 Non-recursive Median Filtering Technique

In non-recursive median filtering technique, the previous 'n' frames are buffered and the median of the buffered frames is employed to construct the background model. The background model is periodically updated for every incoming frame. The foreground pixels are obtained by subtracting the current pixel value from the median value [6] as follows.

$$f_t(x, y) = \begin{cases} foreground; \ |f_t(x, y) - M_t(x, y)| > T \\ background; \ otherwise \end{cases} \quad (8)$$

where T, is the threshold value, $f_t(x, y)$ is the current pixel value, and $M_t(x, y)$ is the median value at time t. In non-recursive median filtering technique, it is necessary to use the buffer with size 'n' in order to store the 'n' previous frames for background modeling. In addition to that, it requires a higher computational cost.

2.3.2 Recursive Median Filtering Technique

In recursive median filtering technique, the median value is incremented or decremented by comparing the incoming pixels instead of buffering the previous frames. If the input pixel is higher than the background pixel, the median is incremented and vice versa. Otherwise, the median value remains constant for the next frame. The background model is updated as follows [7]:

$$Bm_{t+1}(x, y) = \begin{cases} Bm_t(x, y) + 1; \ f_t(x, y) > B_{mt}(x, y) \\ Bm_t(x, y) - 1; \ f_t(x, y) < B_{tm}(x, y) \\ Bm_t(x, y); \ f_t(x, y) = B_{mt}(x, y) \end{cases} \quad (9)$$

where $Bm_{t+1}(x, y)$ is the next background model, $Bm_t(x, y)$ is the current background image frame, and $f_t(x, y)$ is the current foreground image frame. Though the median filtering approach is insensitive to noise and requires fewer computations, it does not predict the objects that suddenly change their state from foreground to background and vice versa.

2.3.3 Single Gaussian Background Model

In single Gaussian background model, mean and covariance of each pixel is modeled as single Gaussian background model and it is updated through all consecutive frames. The background likelihood of each pixel in the current frame is compared with the background pixel. The pixels with vast variation in the background likelihood are classified as foreground. Single Gaussian is not enough to deal with the outdoor environments because of the different statistical characteristics exist for single pixel due to abrupt motions and moving cast shadows.

2.3.4 Gaussian Mixture Model

To improve the single Gaussian method, a mixture of multiple independent distributions is modeled for each pixel in the adaptive Gaussian mixture model. Initially, each pixel is modeled as a mixture of Gaussian distributions and each distribution is assigned a weight. The distributions with large weights represent the background. If any one of the distributions is matched with a current pixel value, then it is classified as a background pixel and the weights of the distribution will be updated as follows [8].

$$w_{k,t} = (1-\alpha)w_{k,t-1} + \alpha\left(Ma_{k,t}\right) \tag{10}$$

where α and k are the learning rate and the number of distributions respectively. The mixture of Gaussians of a current pixel is represented by $Ma_{k,t}$. For matched Gaussian model, $Ma_{k,t}$ is assigned with '1'. Otherwise, it is assigned with '0'. If all 'K' distributions do not match with the current pixel value, then the least probable distribution is replaced by the new distribution with high variance and low prior weight. Due to slow convergence at the initial stage, the Gaussian mixer model is rendered with false detection in case of sudden illumination changes. Moreover, it requires a priori knowledge about the number of Gaussians to be used and an appropriate tuning mechanism for Gaussian parameters. Additionally, these methods involve more computational complexity.

2.3.5 Hidden Markov Model (HMM)

In Hidden Markov model, the intensity variations of each pixel are modeled as different discrete states and each state represents the respective event occurring in the image scene. The hidden Markov model is used to learn the state transition probability between the states. An event, which is difficult to model, can also be represented by the HMM. For example, in baseball event detection, the video shot can be represented as predefined states such as catch, run, and hit. In vehicle tracking, the video scene can be represented as different states such as a car, road, and shadow. Hidden Markov model is also a good alternative for unsupervised background modeling approaches and even rigid events can be easily modeled by this method. But, it uses only low-level visual features and the predefined states will create the ambiguities. Moreover, the arrangements of discrete states and their transition flow are not unique and it is hard to distinct the states under state ambiguities.

2.3.6 Eigen Space Background Subtraction

This method calculates the mean of k sample background images and the covariance matrix for the input frame. Eigenvalue decomposition is applied to the covariance matrix and the background model is constructed by the most descriptive Eigenvectors. All incoming frames are first projected into Eigen space and reconstructed using the projection coefficient and the Eigen vectors. Further, the foreground object is detected by subtracting the reconstructed frame with the Eigen space background model. This method updates the illumination variations through the Eigenvectors and it is comparatively insensitive to background illuminations. Nevertheless, its use is limited to only static backgrounds. It cannot cope with dynamic backgrounds and it is unsuccessful for outdoor environments.

2.3.7 Visual Background Extraction (Vibe) Algorithm

The Vibe algorithm constitutes the background sphere model by a set of pixels randomly observed from the background. The Euclidian distance is calculated between the incoming pixel and the center of the background model and the pixel classification is done based on the Euclidian distance. This method does not require any kind of parameter tuning. The visual background extraction algorithm is stable for illumination changes and camera jitter. This method is random in nature and does not model the background in each frame. So, every time the results will be different if the method is applied repeatedly to the same video data set.

2.4 Optical Flow Method

In optical flow methods, the flow vectors of moving regions are estimated and grouped according to the optical flow characteristics of each pixel such as velocity and direction for foreground motion detection. The stationary object motion is converted into a linear signal trajectory [9]. This method is based on the assumption that the brightness of the respective pixel has smooth variations over the whole image frame. This method can able to detect the moving target even under camera motion. Optical flow method is more sensitive to the noisy environment which leads to poor tracking performance. Moreover, there is a possibility of abrupt background motion changes due to illumination variations, non-stationary backgrounds, and different target appearances. In those cases, the assumption made in optical flow based approaches become invalid and results in poor object detection. In addition to that, the optical flow method involves more computational complexity. Because of these limitations, it is not being preferred mostly for real-time applications.

2.5 *Feature Point Detectors*

In feature-based approaches, the local features such as blobs or point, edge, corners and small patch are used for moving object detection because of its invariability nature under illumination variations and camera motion. The local features are better to handle the scale changes, rotations, and occlusions. Local features are the pattern set up in an image frame and they are distinguishable from its neighbors in terms of texture and color intensity. These features are normally used to find the image correspondence irrespective of occlusion, viewpoint changes and background clutter. Feature detection is used to find the interest feature points which are locally invariant to rotations and scale changes. Feature extraction algorithm computes the descriptors which transform a local pixel neighborhood into a compact vector representation. The commonly used feature detectors are Harris-Stephens corner detector, minimum-Eigen value corner detector, Scale-invariant feature transform (SIFT) detector and Speeded-up Robust Features (SURF) feature detector.

2.6 *Segmentation Based Approaches*

Segmentation based approaches are employed for moving object detection by partitioning the regions with similar characteristics in a scene image. The accuracy of segmentation algorithm decides the object detection results. There are several segmentation approaches.

2.6.1 Mean Shift Algorithm

Initially, the algorithm is provided with a large number of cluster centers that are randomly selected from the given image data and the cluster center is moved through the number of iterations. Mean shift vector is a vector described by the present and previous cluster centers. Mean shift vector is computed iteratively until the cluster center cannot be changed further. Mean shift algorithm necessitates the proper parameter tuning for better segmentation results.

2.6.2 Graph Cut Algorithm

In graph cut algorithm, the image is represented as a graph and the image segmentation problem is regarded as a graph portioning problem. The pixels in the image are considered as vertices. The vertices are partitioned into distinct sub-graphs by trimming the weights of the edges connecting the vertices. The weights are computed according to the similarity in color, texture, and brightness. The total weights of the trimmed edges between the sub-graphs are referred as a cut. In graph cut algorithm, the minimum-cut criterion is affected by the over segmentation. Even though the over-segmentation problem is eliminated by the normalized-cut criterion, it requires more computations and memory storage.

2.6.3 Active Contour

Active contour-based image segmentation describes the object boundaries by the closed contours and these closed contours are evolved based on the image energy function. The fitness of the image contour is characterized by the energy function and this energy function can be computed globally or locally. The local energy computations make use of an image gradient whereas the global energy computations often use the features such as color and texture. Since the image gradient is very sensitive to local minima, the region-based color information is used for image contour evaluation. However, it does not give the good localization of object contour. A major issue with contour based approaches is the contour initialization.

Active contour model approaches need a priori knowledge about the object or background. It is necessary to select the suitable energy function, proper contour initializations and an appropriate contour representation for better segmentation results. Improper energy function, contour representation, and contour initialization will lead to poor segmentation results.

2.7 Supervised Learning Approaches

In supervised learning approach, different object viewpoints are learned customarily from the trained samples for object detection. The learning of different appearance of an object does not claim the complete storage of a set of templates. In the context of object detection, the learning instances are composed of the object features and their associated classes manually. Here, the selection of appropriate object features is crucial because that plays an important role while separating the object class to one another. The other features such as object area, orientation, and density function are also used for learning strategies. There are different learning approaches including decision trees, neural networks, support vector machines and adaptive boosting.

2.7.1 Adaptive Boosting

Adaptive boosting is an iterative method in which the several weak classifiers are combined to pinpoint the accurate classifier in such a way that the performance is better than that of an individual classifier. First, the training set is assigned with initial weights and the errors are estimated by the object classifier. The classifier which gives less classification error is selected as a weak classifier. Then, the weights are increased and the new weak classifier is selected in the next iteration. In perspective of object detection, the weak classifiers are a set of simple thresholds used for the object feature extractions.

2.7.2 Support Vector Machine (SVM)

SVM is a classifier which clusters the image pixels into foregrounds and backgrounds. An image plane, which has the support vectors, is called as a hyper plane. The hyper plane usually separates the image data points into positive and negative samples. The positive samples are referred to as foreground objects and negative samples are referred to as non-foreground objects. There are two categories of SVM classifier namely linear SVM and non-linear SVM. In linear classifier, the hyperplane is derived from different possible hyper planes by linear programming approach. In non-linear classifier, the kernels are used for feature extraction and the kernels transform the non-linear data points into high-dimensional space in which the data points become linearly separable. A more care must be taken while estimating the boundary of hyperplane in linear SVM classifier. Moreover, the selection of right kernel for the object classification is not an easy job in case of non-linear SVM classifier.

2.8 Object Detection Based on Feature Extraction

Visually present local features of an image frame can also be used for foreground feature extraction in moving object detection process. Generally, the local features are distinctive and unique in feature space. The moving objects can be tracked by creating the feature correspondence over the sequence of frames. Since the choice of the right features will have a great impact on the system performance, an appropriate image feature should be used for moving object detection. There are several object features including spectral features, spatial features, and temporal features.

2.8.1 Color Features

The color features are the spectral features used in histogram based approaches. Object color usually consists of two factors such as illuminant and reflectance. In most of the cases, the color features are often chosen because of its high speed of operation, fewer computations and easy to access. Color feature is the most desirable feature in real-time applications such as vehicle tracking, people detection, and human behavior analysis. Among other local features, color information receives much attention due to its consistency under target rotations and scale changes. However, the color features are very sensitive to illumination changes.

2.8.2 Edge Features

Edges are the spatial features used to detect the intensity changes exist around the object boundaries. The edge features are less responsive to illumination variations than the color features. Shape-based object detection method is mostly pursuing the edge features for foreground detection [10]. However, it suffers from ghost problem when stirred object starts moving.

2.8.3 Gradient Features

Gradient refers to the gradual blend of color. Gradient features can be used to extract foreground information in object tracking. In gradient features, each pixel point represents the intensity changes at that point for a given direction. Gradient-based methods use shape or contour to represent the objects, such as the human body [11].

2.8.4 Texture Features

Texture is a process of quantifying the intensity variation of an object in the given image [12]. The texture of an object gives the details about the spatial collection of intensities in the region of interest. It can be found in video frames naturally and it can be used to segment the foreground information. Texture features are customized for illumination changes and shadowing problems. Nevertheless, the texture features do not give sufficient foreground information under smooth surface regions.

2.9 Shape Based Moving Object Detection

Objects can be described by a set of image representations and these representations may be suitable for moving object detection in different occasions [13]. The commonly used object representations are explained below.

2.9.1 Point Representation

In point representation, an object is represented by a single point called Centroid [14]. The object can also be represented by multiple points [15]. This kind of object representations is more appropriate for tracking small moving object regions in a video scene because of its simplicity. Here, the object is represented directly by its pixel intensity or color features.

2.9.2 Primitive Geometric Shapes

Geometric shapes such as rectangle, square, ellipse, and circle are more popular for representing the simple rigid objects in computer vision applications. In such representations, the objects are modeled by translation, affine, or projective transformation. Sometimes, non-rigid objects can also be modeled by primitive geometric shapes [16].

2.9.3 Articulate Shapes

For human behavior analyses the articulated shapes are mostly preferred. In articulate shape models, different parts of the region of interest are joined together by kinetic motion model. Human body composed of body parts such as legs, hands, and a head is an example of an articulated object. Here each part can be represented by primitive shapes including rectangular and ellipse [13].

2.9.4 Skeletal Representation

Skeleton object representation can be used for activity-based object recognition. In this representation, medial axis transform can be applied to object silhouette to extract skeleton of objects. It is also used to model both rigid and articulated objects.

2.9.5 Silhouette Representation

An outline representation of an object consistently filled with black is referred as a silhouette of the object. Silhouette representations are right choice for tracking complex non-rigid objects [13] and it gives the satisfied binary representation for the human motion. However, it is supposed to give the silhouette representation for the non-target object when it is carried by the person.

2.9.6 Active Contour Representation

Active contour representation is an appropriate choice for non-rigid object representation and it describes the object by its edges instead of its entire silhouette [17].

3 Related Works

More contributions have been presented in the literature of moving object detection algorithms. Every one of these endeavours concentrates on a few diverse examinations in the context of object detection and tracking problems. Hence, it is necessary to survey the existing methodologies in different aspects. Table 1 deliberates the detailed review of related works in moving object detection and tracking.

4 Motivation

Today, the essential for an intelligent surveillance system is rapidly growing due to its everlasting applications in different horizons. On the other hand, an enormous video data and the complex surroundings are impaired the surveillance system and it necessitates fast, robust and dedicated automatic moving object detection and tracking algorithms. Several research contributions have been proposed for the moving object detection and tracking. However, robust moving object detection is still a challenging task because of the various complexities involved in context with moving object detection. The visual challenges of moving object detection process are highly motivated the book chapter to conduct certain investigations on the moving object detection algorithms in different complex situations. This book chapter presents three different approaches for moving object detection in order to aid the video surveillance system. The first approach is cross-correlation based novel feature extraction algorithm for moving object detection. This algorithm exploits the foreground features in compressed video domain using cross-correlation and it is insensitive to illumination changes, occlusions and target translations. The second approach is robust skin color based moving object detection. This approach is robust to slowly varying illumination, target rotations, scaling, occlusions, and abrupt motion changes. The third approach is a fusion of color and MD-SILTP texture features algorithm. This algorithm detects the target under varying illuminations, flat surface regions, partial occlusions, scaling, dynamic backgrounds, and moving cast shadow. The keen interest of these research methods is to detect and locate the moving objects such as toys, human face, and a car in complex scenes without background interventions.

Table 1 Summary of the related works

Methods	Author	Description	Limitations
Background subtraction and background modeling	Bhaskar et al. [18]	Dynamic inverse analysis is used for background modeling and thereby updates the dynamic background variations	It is not able to update the rapid variations
	Barnich and Droogenbroeck [19]	The background sphere model is constituted with a set of pixels observed from the background randomly	Results will be inconsistent
	Sheikh et al. [20]	Proposed background subtraction method to extract the foreground object for motion analysis	Not able to handle with background changes
Optical flow methods	Jansari and Parmar [21]	Optical flow vectors are used for moving object detection	More sensitive to noise and occlusion conditions
	Aslani and Mahdavi-Nasab [22]	Use the Horn-Schunck algorithm for the traffic monitoring surveillance system	Camera shaking and object occlusion impaired the system performance
Texture and color features based approach	Liao et al. [23]	Scale Invariant Local Ternary Patterns (SILTP) operator is proposed	Inefficient for noisy image frames Cannot work robustly in the flat and dark areas
	Lee et al. [24]	Opponent Color LBP (OCLBP) to handle the illumination variations and smooth surface regions	Easily affected by illumination noises More computational complexity
	Zhang et al. [25]	Multi-scale fusion strategy is used in which each pixel is encoded with N different binary patterns	The feature vectors become too sparse and it loses few foreground parts
	Ma and Sang [26]	Multichannel-SILTP (MC-SILTP) is proposed	Spatial non-saliency Sensitive in featureless areas

(continued)

Table 1 (continued)

Methods	Author	Description	Limitations
	Han et al. [27]	Fusion of SILTP and color is proposed and dominant background patterns are use	Foreground patterns are wrongly considered as background patterns
	Narayana et al. [28]	Multiple variances are calculated and the best variance is selected for updating the dynamic changes	Shadows are being mistakenly classified as foreground
Feature extraction techniques	Babenko et al. [29]	Multiple instance learning approach (MIL) is used	Not suitable for large scale changes
	Mei and Ling [30]	L1-minimization (L1Track) is used for object modeling	Computational complexity is high
	Zhang et al. [31] (FCT)	Most discriminative features are extracted through data-independent sparse measurement matrix	Does not work under heavy Occlusion
	Fan et al. [32] (OLDD)	It incorporates a priori class label information into the dictionary learning process	It requires prior-knowledge about the label information

5 System Overview

Despite the fact that the moving object detection approaches have been studied for many years, still it remains an open research problem and it is a great challenge to achieve robust, precise and superior methodology. For example, it is more difficult to identify a particular human face under heavy illumination changes and different poses. In practice, no single algorithm deal with all challenges in moving object detection and tracking. In this section, different novel approaches are presented to detect and locate the targets including car, toy, and a human face in complex scenes without background influences. The methods are as follows:

- A feature extraction algorithm based on cross-correlation coefficients is presented to deal with target translations, illumination changes, and partial occlusions.
- A robust skin color based moving object detection algorithm is proposed to cope with the situations that the in-plane/out-of-plane rotations, motion blurring, scale changes, partial occlusions, slowly varying illumination conditions and abrupt motion changes.

- The fusion based approach, in which texture and color features are fused at the decision level, is proposed to tackle with abrupt illumination changes, similar texture, abrupt motion changes, scale changes, moving cast shadow and dynamic background changes (i.e., waving tree and curtains).

5.1 Feature Extraction Based on 2D Cross-Correlation

The scope of this method is to detect the targets such as toy and a human face under illumination changes, object translations, and target occlusions. To overcome the shortfalls of the existing techniques, a novel feature extraction algorithm has been proposed in this section. This feature extraction algorithm adapts the cross-correlation to derive the matched feature points in the given video frame. Since the cross-correlation does not need to search the parameters, it provides a simple feature extraction method for moving object detection. Figure 3 shows the overview of 2D cross-correlation based feature extraction. Initially, the input video frames are compressed using the 2D discrete cosine transform. Since the 2D-DCT has highest energy compaction, it gives the significant compression results and thereby increasing the processing speed. Then, the similarity measure is performed between the compressed video frame and the target template by estimating the cross-correlation coefficients. The maximum likelihood densities for the pixels being foreground are formulated by estimating the peak correlation coefficients. In addition to that, the maximum likelihood densities for the pixels being background are also formulated from the rest of the feature points. The features points are classified into foreground and background based on the Bayesian classification rule. Finally, the extracted foreground features are tracked in the successive image frames by calculating the bounding box around the foreground objects. Furthermore, the peak-to-side lobe ratio (PSR) is calculated for effective handling the occlusion conditions. The PSR drops to zero when the target is completely occluded. Under such circumstances, the target model is derived based on the previous target information.

This approach effectively handles the challenging environments including target translations and partial or complete occlusions as compared to the other baseline methods. However, in view of rapid variations on both camera and target under dynamic environments, the target information is not enough for accurate object detection. Hence, this approach does not perform well in dynamic backgrounds. The 2D cross correlation is used for feature extraction to detect the presence of the object in the given video frames. It is sensitive to the image rotation and scaling which degrade the tracking performance. Moreover, the target which is stationary for a long time in video sequence misleads the object tracker into false detections.

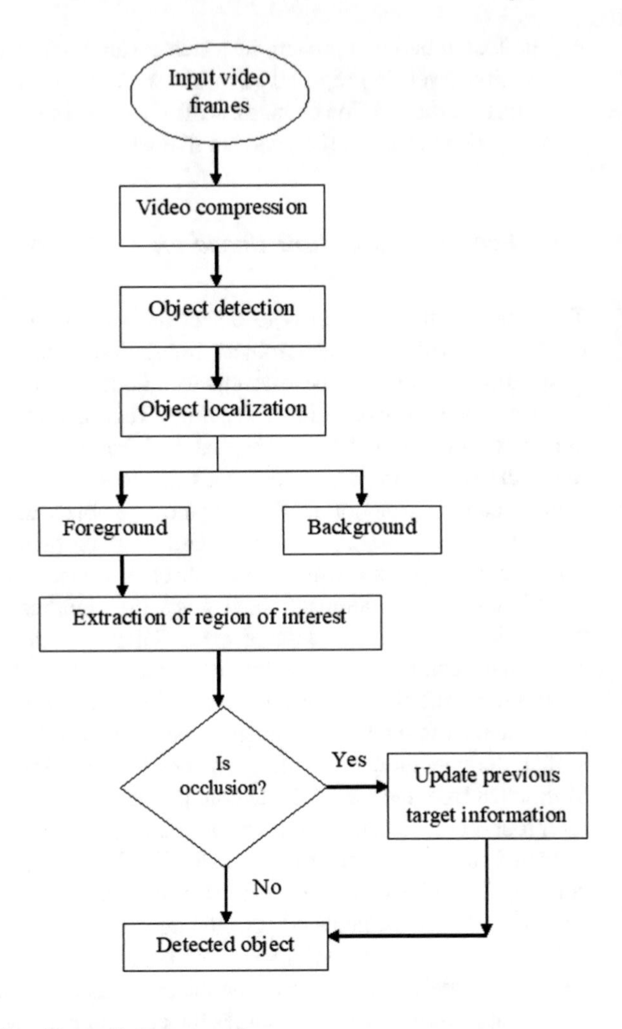

Fig. 3 Flow diagram of feature extraction based on 2D cross correlation scheme

5.2 *Robust Skin Color Based Approach*

The overview of skin color based approach is shown in Fig. 4. The main objective of this method is to identify and track the human face in under target rotations/scaling, abrupt motion changes, pose variations, cluttered backgrounds, and occlusions without background influences. In the first stage, the input image frame is smoothed using an averaging filter in order to reduce the nonlinear effects of poor illumination conditions. Then, the smoothed RGB image frame is converted into YCbCr color space. Here, YCbCr color space is selected because of its explicit separation between luminance and chrominance components, and the chrominance components are mostly unaffected by the illumination changes. In the second stage, the skin color region is detected by applying Otsu's method of gray level thresholding [33] in each of the three components of YCbCr image frame. Then, the skin color

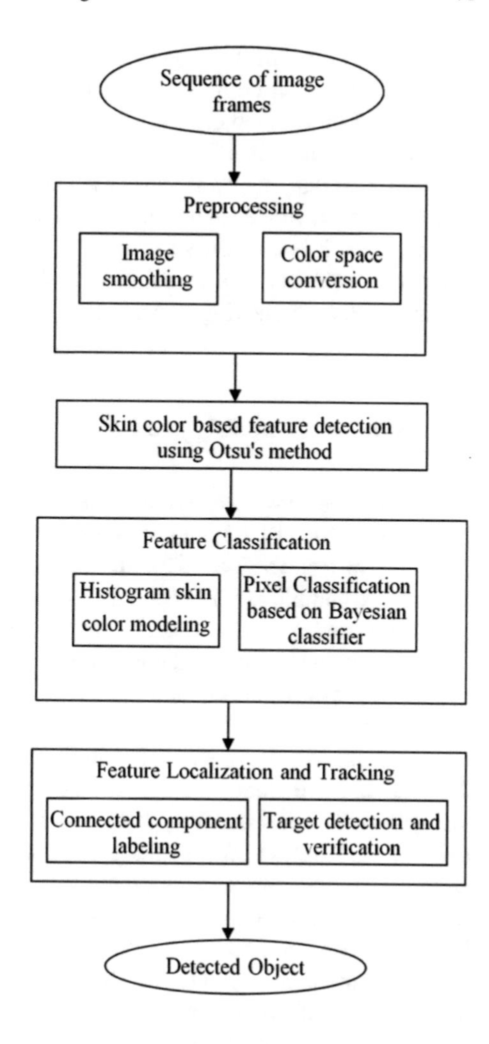

Fig. 4 Overview of skin color based approach

region is identified by combining the individual threshold binary image frames using Boolean AND operation. In the third stage, the skin and nonskin pixels are extracted based on the thresholding results obtained in the second stage. Then, skin and nonskin color histograms are estimated using skin and nonskin pixels. The histogram-based Bayesian skin classifier is constructed and the skin color regions are classified into foreground and background. This approach samples the skin pixels according to the threshold binary image and constructs the skin and nonskin histograms. In the next stage, the skin color features are localized by the label-equivalence-based connected component labeling process and it is based on the eight-connectivity neighboring analysis. Finally, the localized foreground skin regions are confirmed as target (i.e., face) regions by verifying the region properties followed by Euler method, and it is tracked by placing the bounding box around the target in successive video frames. The skin color-based method can successfully detect the target without background influences in most of the sequences and the experiment results clearly mention the novelty of the skin color approach under complex situations. However, the skin color based moving object detection approach mainly depends on the single skin color features and it is failed to detect the target under rapid illumination variations. In addition to that, the skin color-based approach involves global threshold for skin color detection. Since global thresholding is easily influenced by the background disturbances, it is necessary to compute the local adaptive threshold for skin color detection. Moreover, it is not suitable for dynamic backgrounds such as waving tree, rippling water and moving curtains etc.

5.3 Fusion Based Approach

To alleviate the above issues, the fusion of median based scale invariant local ternary pattern (MD-SILTP) texture and color features is used for moving object detection. Here, the color and MD-SILTP texture features are extracted independently and those features are combined at the feature classification level. Figure 5 shows the overview of fusion based approach. First, the input image frame is converted into a grayscale image frame and it is divided into several sub-regions. Then, the MD-SILTP texture patterns are estimated for each local region. After that, the MD-SILTP histograms are constructed and matched with the background model histogram through the histogram intersection method. Similarly, the color histograms are constructed from the original RGB image frames and they are matched with the background model histograms. The best matching color and texture model histograms are updated with the new background information by updating its bins. At the end, the MD-SILTP texture and color features are combined for final feature classification. The color features provide good supplement information under the flat surface regions and the MD-SILTP texture features give the useful information about the target in case of varying illumination conditions. Therefore, this kind of approach is able to address both the flat surface regions and varying illumination conditions in a better way as compared to the others. In addition to that, the MD-SILTP does not consider the inter-channel differences and it is more significant than the other baseline methods.

6 Performance Evaluation

This section presents the performance evaluation of all the three proposed works. The performance metrics used for the analysis are described and finally, a detailed review of the results are given at the end of the section.

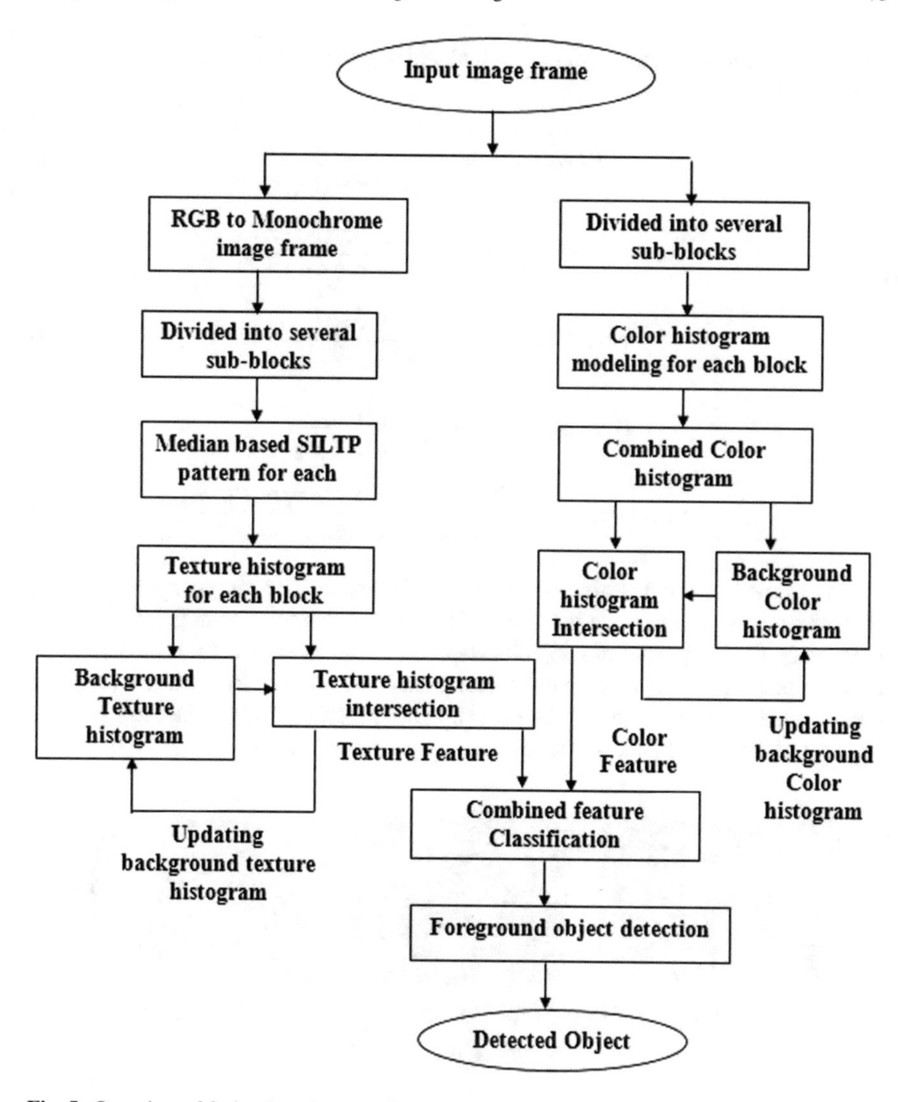

Fig. 5 Overview of fusion based approach

6.1 Evaluation Metrics

Several frame based evaluation metrics are available in the literature to assess the performance of the moving object detection algorithms. The common metrics such as precision, detection rate, f-measure, false alarm rate, accuracy are used for the performance analysis of the proposed approaches. All frame based metrics are estimated using the frame based constraints such as true positive (T_{pos}), true negative (T_{Neg}), false positive (F_{pos}) and false negative (F_{Neg}) [34, 35]. For all testing sequences, the frame based-surveillance metrics are calculated based on the frame based-constraints as follows:

$$Precision = \frac{T_{pos}}{\left(T_{pos} + F_{pos}\right)} \quad (11)$$

$$Accuracy = \frac{\left(T_{pos} + T_{Neg}\right)}{\left(T_{pos} + F_{pos} + T_{Neg} + F_{Neg}\right)} \quad (12)$$

$$FalseAlarmRate = \frac{F_{Pos}}{(T_{Pos} + F_{Pos})} \quad (13)$$

$$DetectionRate = \frac{T_{pos}}{\left(T_{pos} + F_{Neg}\right)} \quad (14)$$

$$F\text{-}measure = \frac{2 \times Precision \times DetectionRate}{(Precision + DetectionRate)} \quad (15)$$

The frame based constraints are illustrated in Fig. 6. The sample calculations of frame based constraints are given in Fig. 7a–f. The rectangular bounding box is used to identify the foreground moving objects in the given image frame. If the bounding box region contains the foreground object, then it is said to be true positive (T_{pos}). If the object is not present inside the bounding box, then it is said to be false positive (F_{pos}). On the other hand, the target missed by the bounding box is referred as false

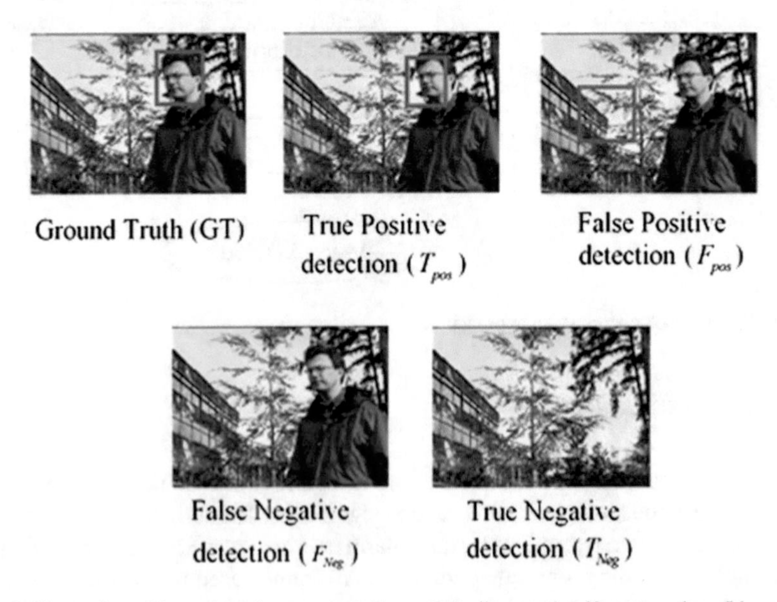

Fig. 6 Illustration of frame based constraints. *Source* http://perception.i2r.a-star.edu.sg/bk_model/bk_index.html [3]

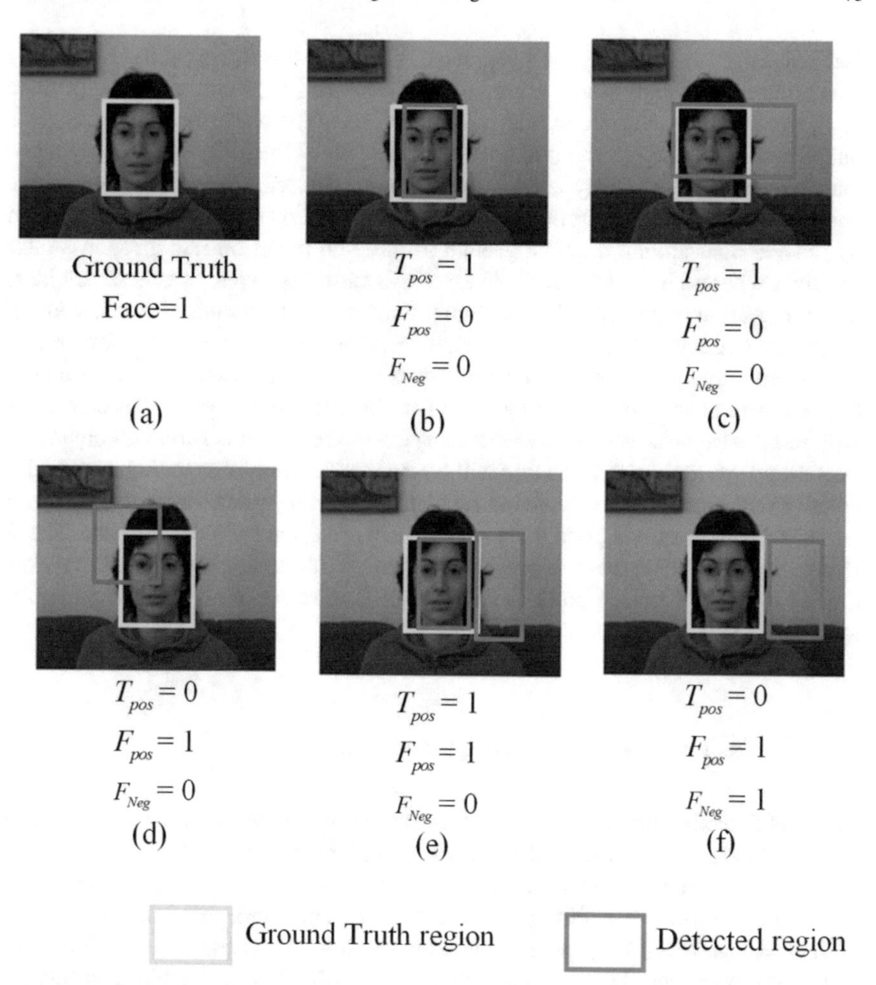

Fig. 7 Sample calculation of frame based constraints such as true positives (T_{pos}), false positives (F_{pos}), true negatives (T_{Neg}) and false negatives: **a** illustration of ground truth region, **b–f** method of calculating the frame based constraints. *Source* http://cvlab.hanyang.ac.kr/tracker_benchmark/datasets.html [36]

negative (F_{Neg}) and the correct rejection of non-foreground objects is termed as true negatives (T_{Neg}).

The ground truth bounding box region is compared with the detected bounding box regions in order to validate the numerical results. The frame based constraints are calculated based on the overlapping between ground truth region and detected region. In proposed approach, the overlapping threshold is set to 0.5 for correct detections.

For example, in Fig. 7b, c, the detected region overlaps with ground truth region by more than 50% and it gives one true positive (T_{pos}). In Fig. 7d, the detected region overlaps with ground truth region by less than 50% and it gives zero true positives and one false positive. On the other hand, detecting the foreground as well as a background in a single frame gives one true positive (T_{pos}) and one false positive (F_{pos}) as shown in Fig. 7e. Similarly, only detecting the background regions gives one false positive (F_{pos}) and one false negative (F_{Neg}) (See Fig. 7f). The precision metric measures how many labeled positives are correct among the total positive detections, whereas the detection rate measures how many true positives are selected by the classifier among the total ground truth. High precision rate gives less false positives whereas high detection rate gives less false negatives. In general, an ideal object detection approach achieves high precision and high detection rate with all pixels are correctly classified. Conversely, a practical system predicts foreground pixels with more or fewer classification errors and it is not possible to achieve both high precision and high detection rate. If the classifier allows number of detections with more false positives and less false negatives, then the precision will obviously be reduced while the detection rate will be increased. In addition to that, more false positive will increase the false alarm rate and reduce the system accuracy. Hence, it is necessary to achieve a good trade-off between precision and detection rate. The f-measure is a harmonic mean of precision and recall. Increasing the f-measure makes a good trade-off between precision and recall. The range of all the frame metrics must lie between 0 and 1.

6.2 Novelty of the Proposed Approaches

In object detection process, the classifiers should minimize the errors such as false negatives and false positives. The object classifier which creates fewer false negatives is mostly preferred even if it creates more false positives. This is so because there is a chance to correct the false positive errors in a later stage, but it is not possible with false negatives. Moreover, it is far better to allow occasional false positives than to fail to notice real objects. Therefore, the main objective of the object detection algorithm is to minimize the false negative rate by accepting comparatively high false positive rates. For comparative analysis, all the proposed approaches are tested on the challenging sequences including FaceOcc1, David Indoor, shaking2, and curtains. The David indoor sequence is used to evaluate the performance under illumination variation, in-plane/out-of-plane pose change, and partial occlusion. The FaceOcc1 sequence is used to evaluate the performance under in-plane pose variations, and partial occlusions. The Shaking2 sequence is used to test the performance under fast motion, motion blurring, in-plane/out-of-plane rotations.. The challenging sequences such as waving tree and curtain sequences are used to test the performance against the dynamic backgrounds. The performances are evaluated based on the frame based metrics such as false alarm rate, accuracy, precision, detection rate and F-measure. From the numerical results, it is noticed that the cross-correlation based feature extraction algorithm achieves an average false alarm rate of 8.93%, accuracy of 90.69%, precision rate of 91.07%, and occlusion rate of 58.49% for the selected video sequences. This algorithm provides the solution for partial/complete occlusions, illumination changes, and target translations. However, the cross-correlation coefficients are highly sensitive to target rotations, abrupt motion changes, motion blurring and scaling. The skin color based approach gives an average centre location error of about 16.97, success rate of 86%, precision rate of 84.21%, accuracy of 85.48%, false alarm rate of 15.79%, and detection rate of 86.88% for the selected sequences. This method addresses the issues such as target scale changes, target rotations, target motion changes and target occlusions. The fusion based MD-SILTP method achieves the average precision rate of 82.8%, detection rate of 80.4%, F-measure of 81.6%, false alarm rate of 17.2%, and Success rate of 81.4%. Low false alarm rate indicates that the proposed algorithms effectively resist the background disturbances. The higher values of precision, accuracy, detection rate, and F-measure and success rate indicate that the proposed approaches perform better than the other baseline methods under complex situations.

7 Conclusion and Future Directions

In summary, the novel approaches presented in this chapter give more promising results under complex environments. Nevertheless, there is still a space to improve the system performance. Particularly, dynamic background handling, multiple object tracking, and processing time are the most promising subjects to improve further for robust moving object detection and tracking. The correlation based approach is not able to deal with target rotations, scaling and dynamic variations of background. The estimation of cross-correlation coefficient demands high computational complexity and do not give the target details under scaling and rotations. The sophisticated motion estimation algorithms can help to improve this algorithm and online processing feature extraction algorithm can serve as a basis to model and update the dynamic background changes. Robust skin color based approach relies only on the skin color features and these features are more sensitive to illumination changes. Hence, it is necessary to use additional local features to cope with the rapid illumination changes. In addition to that, the proposed approach is not applicable to on-line processing and it may be extended further by reducing the computational complexity of skin histograms. The MD-SILTP based moving object detection is computationally intensive. Further, a single histogram is not enough to deal with multimodal dynamic background variations. Though the color features are incorporated to deal with smooth surface regions, the algorithm still needs advancement for better handling the texture-less regions and abrupt motion changes. Further, it is not suitable for online processing. The real time characteristics depend on the frame rate, computational cost of the tracker, and the computing capability of the hardware. The proposed methods require high computational complexity and working in low frame rate. Hence, these approaches are not applicable to online processing. At the end, it is concluded that the book chapter has contributed the new ideas to handle the major challenges in the context of moving object detection and the system performance may further be improved with real time applications in the computer vision.

References

1. http://cvlab.hanyang.ac.kr/tracker_bench
2. http://www4.comp.polyu.edu.hk/~cslzhang/FCT/FCT.html
3. http://perception.i2r.a-star.edu.sg/bk_model/bk_index.html
4. Rout RK (2013) A survey on object detection and tracking algorithms
5. Zhang LZL, Liang YLY (2010) Motion human detection based on background subtraction. In: 2010 Second international workshop on education technology and computer science (ETCS), vol 1, pp 284–287
6. Lo BPL, Velastin SA (2001) Automatic congestion detection system for underground platforms. In: Proceedings of 2001 international symposium on intelligent multimedia, video and speech processing, ISIMP 2001 (IEEE Cat. No.01EX489), pp 158–161
7. Parks DH, Fels SS (2008) Evaluation of background subtraction algorithms with post-processing. In: Proceedings—IEEE 5th international conference on advanced video and signal based surveillance, AVSS 2008, pp 192–199
8. Stauffer C, Grimson WEL (2000) Learning patterns of activity using real time tracking. IEEE Trans Pattern Anal Mach Intell 22(8):747–757
9. Kinoshita K, Enokidani M, Izumida M, Murakami K (2006) Tracking of a moving object using one-dimensional optical flow with a rotating observer. In: 2006 9th International conference on control, automation, robotics and vision
10. Bowyer K (2001) Edge detector evaluation using empirical ROC curves. Comput Vis Image Underst 84(1):77–103
11. Deori B, Thounaojam DM (2014) A survey on moving object tracking in video. Int J Inf Theor (IJIT) 3(3):31–46
12. Yang H, Shao L, Zheng F, Wang L, Song Z (2011) Recent advances and trends in visual tracking: a review. Neurocomputing 74(18):3823–3831
13. Yilmaz A, Javed O, Shah M (2006) Object tracking: a survey. ACM Comput Surv 38(4):1–43
14. Veenman CJ, Reinders MJ, Backer E (2001) Resolving motion correspondence for densely moving points. IEEE Trans Pattern Anal Mach Intell 23(1):54–72
15. Serby D, Gool LV (2004) Probabilistic object tracking using multiple features. In: Proceedings of 17th International conference on the pattern recognition (ICPR'04) vol 2, pp 184–187
16. Comaniciu D, Ramesh V, Meer P (2003) Kernel-based object tracking. IEEE Trans Pattern Anal Mach Intell 25(5):564–577
17. Li, X, Hu, W, Shen, C, Zhang, Z, Dick, A, Hengel, AVD (2013) A survey of appearance models in visual object tracking. ACM Trans Intell Syst Technol 4(4):58:51–58:48
18. Bhaskar H, Dwivedi K, Dogra DP, Al-Mualla M, Mihaylova L (2015) Autonomous detection and tracking under illumination changes, occlusions and moving camera. Signal Processing 117:343–354
19. Barnich O, Droogenbroeck MV (2011) ViBe: a universal background subtraction algorithm for video sequences ViBe: a universal background subtraction algorithm for video sequences. IEE Trans Image Process 20(6):1709–1724
20. Sheikh Y, Javed O, Kanade T (2009) 'Background subtraction for freely moving cameras. In: Proceedings of IEEE 12th international conference on computer vision, pp 1219–1225
21. Jansari D, Parmar S (2013) Novel object detection method based on optical flow. In: Proceedings international conference on emerging trends in computer and image processing, Malaysia, pp 197–201, 8–9 Jan 2013
22. Aslani S, Mahdavi-Nasab H (2013) Optical flow based moving object detection and tracking for traffic surveillance. Int J Electr Comput Energ Electron Commun Eng 7(9):963–967

23. Liao S, Zhao G, Kellokumpu V, Pietikäinen M, Li SZ (2010) Modeling pixel process with scale invariant local patterns for background subtraction in complex scenes. In: Proceedings of the IEEE computer society conference on computer vision and pattern recognition, pp 1301–1306
24. Lee Y, Jung J, Kweon I-S (2011) Hierarchical on-line boosting based background subtraction. In: Proceedings of 17th Korea-Japan joint workshop on frontiers of computer vision (FCV), pp 1–5
25. Zhang Z, Wang C, Xiao B, Liu S, Zhou W (2012) Multi-scale fusion of texture and color for background modeling. In: Proceedings—2012 IEEE 9th international conference on advanced video and signal-based surveillance, AVSS 2012, pp 154–159
26. Ma F, Sang N (2013) Background subtraction based on multi-channel SILTP. In: Lecture notes in computer science (including subseries Lecture Notes in Artificial Intelligence and Lecture Notes in Bioinformatics), vol 7728 LNCS, no PART 1, pp 73–84
27. Han H, Zhu J, Liao S, Lei Z, Li SZ (2015) Moving object detection revisited: speed and robustness. IEEE Trans Circ Syst Video Technol 25(6):910–921
28. Narayana M, Hanson A, Learned-Miller E (2012) Background modeling using adaptive pixelwise kernel variances in a hybrid feature space, pp 2104–2111
29. Babenko B, Yang MH, Belongie S (2011) Robust object tracking with online multiple instance learning. IEEE Trans Pattern Anal Mach Intell 33(8):1619–1632
30. Mei X, Ling H (2011) Robust visual tracking and vehicle classification via sparse representation. IEEE Trans Pattern Anal Mach Intell 33(11):2259–2272
31. Zhang K, Zhang L, Yang M-H (2014) Fast compressive tracking. IEEE Trans Pattern Anal Mach Intell 36(10):2002–2015
32. Fan B, Du Y, Cong Y (2014) Online learning discriminative dictionary with label information for robust object tracking. In: Abstract and applied analysis, vol 2014
33. Liao P-S, Chen T-S, Chung P-C (2001) A fast algorithm for multilevel thresholding. J Inf Sci Eng 17(5):713–727
34. Fawcett T (2006) An introduction to ROC analysis. Pattern Recogn Lett 27:861–874
35. Powers DMW (2011) Evaluation: from precision, recall and F-factor to ROC, informedness, markedness & correlation
36. http://cvlab.hanyang.ac.kr/tracker_benchmark/datasets.html

Enhanced End-to-End System for Autonomous Driving Using Deep Convolutional Networks

Balaji Muthazhagan and Suriya Sundaramoorthy

Abstract The emergence of autonomous cars in today's world makes it imperative to develop superlative steering algorithms. Deep convolutional neural networks are widely adopted in vision problems for their adept nature to classify images. End-to-end models have acted as an excellent substitute for handcrafted feature extraction. This chapter's proposed system, which comprises of steering angle prediction, road detection, road centering, and object detection, is a facilitated version of an autonomous steering system over just considering a single-blind end-to-end architecture. The benefits of proposing such an algorithm for the makeover of existing cars include reduced costs, increased safety, and increased mobility.

Keywords Autonomous vehicles · Self-driving car · End-to-End deep learning · Convolutional neural networks · Road detection · Object detection

1 Introduction

Autonomous driving refers to an action implemented by autonomous driving systems (ADS) which can steer or guide itself by perceiving the inputs of its environment without human intervention [1]. Companies such as Tesla, Waymo, Volkswagen, Hyundai etc. are investing extensively in the research and development of such systems [2]. Apart from consumer-oriented vehicles, these types of systems are largely being probed whether they can be a good fit for trucks which cover large distances [3, 4].

There are six levels of automation which are defined by the SAE (Society of Automotive Engineers) in the J3016™ standard [5]:

B. Muthazhagan (✉) · S. Sundaramoorthy
PSG College of Technology, Coimbatore, Tamil Nadu 641004, India
e-mail: balajimuthazhagan@gmail.com

S. Sundaramoorthy
e-mail: suriyas84@gmail.com

K. R. Ahmed et al. (eds.), *Deep Learning and Big Data for Intelligent Transportation*, Studies in Computational Intelligence 945,
https://doi.org/10.1007/978-3-030-65661-4_4

- Level zero: In this level automation does not contribute to the driving of the vehicle. The driving is completely controlled by the human driver. The system can involve in the issuance of cautions like the emergency braking system but does not assist in driving.
- Level one (Hands on): This is the level in which automation is introduced. The driver is supported with the help of the system in place for fine steering. The steering of the vehicle is shared between the driver and the system. Adaptive cruise control where the vehicle automatically adjusts the steering based on the distance of the vehicle in front is an example of this level.
- Level two (Hands off): In this level, the steering is completely controlled by the underlying system. However, continuous monitoring of the vehicle's actions is necessary. The term hands off here does not explicitly mean that the driver can take their hands off the steering. SAE has in fact made it mandatory to keep the hands on the steering wheel while the system is driving. Tesla's Full Self-Driving Autopilot system is an example of this level.
- Level three (Eyes off): In this level, the driver need not focus on the road and the autonomous vehicle will handle the complete steering. But in the event of emergency situations, the driver is still needed to intervene and take control.
- Level four (Mind off): In the event of emergencies, level four autonomous systems can take over and suggest actions. Manual override is also possible from the driver. Current legislations allow level four vehicles only in a geofenced area or in cases such as traffic jams citing safety. Google's Waymo project which is geofenced to regions in the US is an example of level four autonomy.
- Level five (No steering wheel): The driver is completely removed from the equation and is no longer required to monitor the steering. Level five vehicles are not be geofenced or restricted to a region.

Autonomous steering is said to be comprised of 3 units [6, 7]:

- Perception—Deals with how the vehicle can receive the visual and other essential inputs. This unit includes the various sensors such as camera, LIDAR, Radar etc.
- Computing—Deals with how the visual input must be processed in order to gain insightful information. This unit comprises of the algorithm engine which is responsible for initiating a steering action.
- Control—Deals with the mechanical aspect of steering the vehicle post the computation stage: once the engine has initiated an action, an actual actuator needs to execute it (Fig. 1).

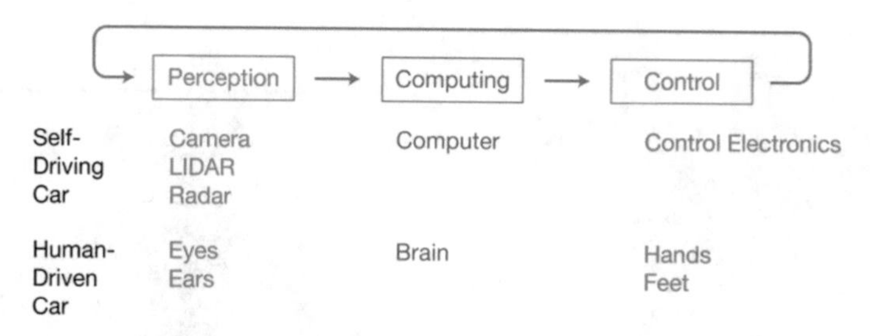

Fig. 1 Breakdown of autonomous and human steering systems

2 Previous Deep Learning Methods for Autonomous Driving

2.1 Alvinn

The first approach to experiment with neural networks for autonomous vehicle navigation was proposed by Pomerleau in 1989 through ALVINN (Autonomous Land Vehicle in a Neural Network) [8] which was designed to send steering commands to NAVLAB, the test vehicle which was used by Carnegie Melon University. A grayscale image of dimensions 30 × 32 was used as the input to this neural network and there were 30 different outputs which were possible. A synthetic road generator was used to generate the training examples over actual road images given the logistic difficulties of capturing the wide variety of training examples. A camera was housed on the top of NAVLAB and real time prediction was carried out. The system was said to accurately drive up to speeds of 2 km/h for 400 m under proper lighting conditions (Fig. 2).

2.2 DAVE

In 2004 the Defense Advanced Research Projects Agency (DARPA) initiated an experiment in which a convolutional neural network-based approach was proposed to guide a radio-controlled truck which was 50 cm in length [9]. The architecture consisted of two convolutional layers alternated by two pooling layers followed by two fully connected layers. There were two cameras, one on the right and one on the left which produced YUV formats of resolution 149 × 58. This was used as the input to the architecture. A human supervisor controlled the vehicle across varied environments and lighting conditions as a part of the training data collection task. The model was trained using 95,000 frames in an end-to-end supervised manner and 31,800 frames were used for testing (Fig. 3).

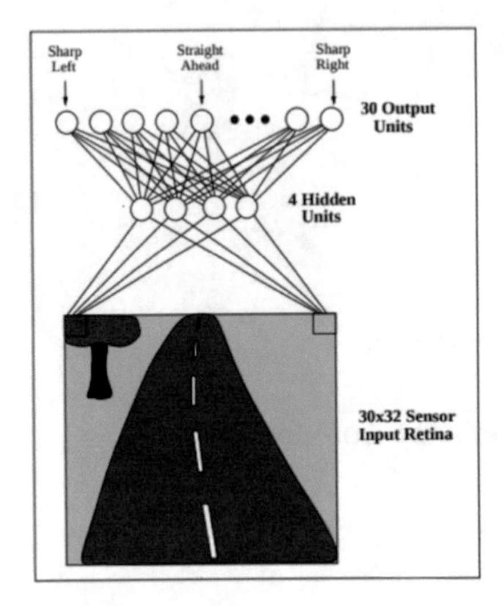

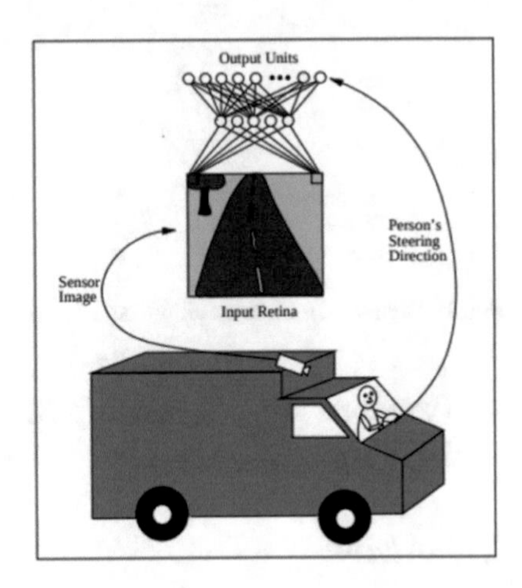

Fig. 2 Neural network architecture and setup on ALVINN

Fig. 3 DAVE robot

2.3 DAVE-2

In 2016 Nvidia implemented a successor to the DAVE project by proposing an end-to-end approach [10]. The architecture of DAVE-2 is deeper than that of DAVE, where five convolutional layers were used instead of two. DAVE-2 enjoyed a good training accuracy because of two main reasons: the multitude of labeled training datasets which were available and the exploitation of GPUs for training CNNs. The input to this architecture is captured from a camera which first undergoes a phase shift and rotation, and data augmentation is adopted to prevent the model from failure

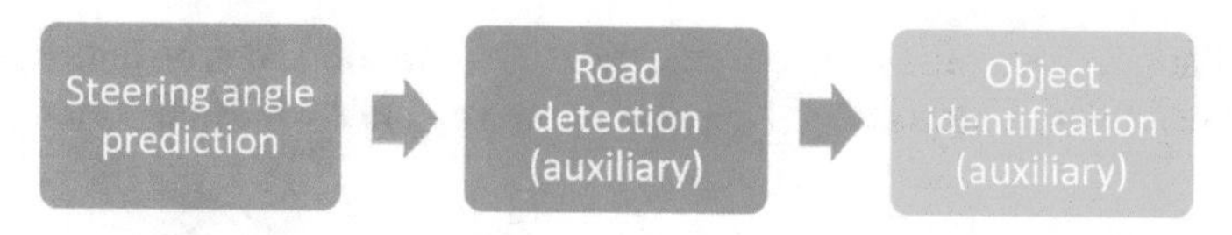

Fig. 4 Training methodology for DAVE-2 [10]

when put before an event which it has not seen before. The actual steering angle was captured with the help of a human driver and is used to train the network. We will also consider the implementation of this architecture in this chapter owing to its good accuracy (Fig. 4).

3 Proposed System

The proposed system takes a live feed or a video as an input and outputs the steering angle, radius of curvature of the road, and objects present in the view with their velocities. Using this information, it is easy to derive what should be steering action and the acceleration that can be safely done (Fig. 5).

The processing pipeline consists of the following components:

- Main component—Steering angle prediction: This component is responsible for the prediction of the steering angle of the vehicle based on the windshield view as input using an end-to-end CNN model.
- Auxiliary component 1—Road detection: This component is responsible for the identification of road markings, and thereby identifying the radius of curvature to steer correctly. This helps us in identifying the false steering angles predicted by the end-to-end component.

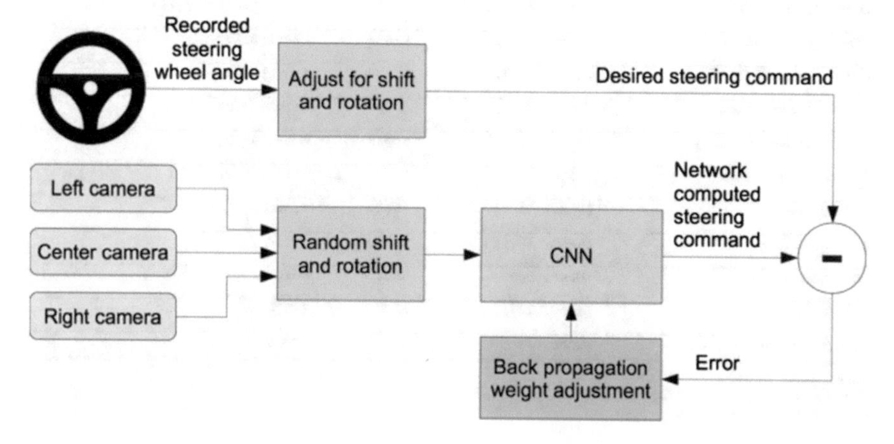

Fig. 5 Proposed system pipeline

- Auxiliary component 2—Object detection: This component is responsible for the identification of objects which appears in the vehicle's windshield view.

3.1 Prediction of Steering Angle

Instead of extracting the features and then passing it on to the prediction phase, end-to-end models take an input and give a corresponding output based on a single learning algorithm. This reduces the error involved in hand-engineered pipelines where features which are deemed important are chosen by humans. The prediction of the steering angle happens with an end-to-end CNN model in which the windshield view is given as the input and the steering angle is taken as the output. In this chapter we consider two popular models to achieve this task.

DAVE-2 CNN model. The model consists of 5 convolutional layers, 1 normalization layer and 3 fully connected layers [10]. The starting layer of the model is responsible for normalization, post which there are convolutional layers which uses strided convolutions having kernel dimensions 2×2 and 5×5 followed by a unstrided convolution of kernel dimensions 3×3 which is used to extract the relevant features. This is followed by fully conventional layers which result in the output. The implementation is done using Keras with ReLU as the activation function. The input images are resized to $64 \times 64 \times 3$ (Fig. 6).

Comma.ai CNN model. The model consists of 3 convolutional layers, 1 normalization layer and 2 fully connected layers [11]. The starting layer of the model is responsible for normalization post which there are convolutional layers which uses strided convolutions having kernel dimensions 8×8 and 5×5 followed by fully

Layer (type)	Output Shape	Param #	Connected to
Normalization (Lambda)	(None, 64, 64, 3)	0	lambda_input_1[0][0]
Conv1 (Convolution2D)	(None, 30, 30, 24)	1824	Normalization[0][0]
Conv2 (Convolution2D)	(None, 13, 13, 36)	21636	Conv1[0][0]
Conv3 (Convolution2D)	(None, 5, 5, 48)	43248	Conv2[0][0]
Conv4 (Convolution2D)	(None, 3, 3, 64)	27712	Conv3[0][0]
Conv5 (Convolution2D)	(None, 1, 1, 64)	36928	Conv4[0][0]
flatten_1 (Flatten)	(None, 64)	0	Conv5[0][0]
FC1 (Dense)	(None, 1164)	75660	flatten_1[0][0]
FC2 (Dense)	(None, 100)	116500	FC1[0][0]
FC3 (Dense)	(None, 50)	5050	FC2[0][0]
FC4 (Dense)	(None, 10)	510	FC3[0][0]
output (Dense)	(None, 1)	11	FC4[0][0]

Total params: 329,079
Trainable params: 329,079
Non-trainable params: 0

Fig. 6 Keras implementation of DAVE-2 model

Fig. 7 Keras implementation of Comma.ai CNN model

conventional layers which result in the output. The implementation is done using Keras with ELU as the activation function. The input images are resized to 64 × 64 × 3.

Two datasets from the region around Rancho Palos Verdes and San Pedro regions around California are considered for the training and validation of these models [12]. The dataset contains the file images and the steering angles associated with them. The datasets were split in as 70:15:15 for training:validation:testing ratio. Also, minimal data augmentation using flips and cropping of the top of the image is used. This eliminates the bias towards a direction in the images (Figs. 7 and 8; Table 1).

Both the models were run on the previously mentioned datasets with the loss function being RMSE along with Adam optimizer to optimize the loss (Table 2).

3.2 Road Detection Using Image Processing

Before we delve into the pipeline implementation of identifying roads, it is necessary to understand three main techniques:

- First is to calibrate the camera
- Second is to solve for the projective transform
- Third is to map pixels in the image frame to distance metrics like meters etc.

Camera calibration. The calibration of the camera [13] is done so as to extract the relevant parameters of the camera used to capture the images since we need to get an undistorted version of the image, which is easier to identify by mapping a three-dimensional point in the real world onto a two-dimensional frame. In other words,

```
Model = comma_ai
____________________________________________________________________________________________________
Layer (type)                     Output Shape          Param #     Connected to
====================================================================================================
Normalization (Lambda)           (None, 64, 64, 3)     0           lambda_input_2[0][0]
____________________________________________________________________________________________________
conv1 (Convolution2D)            (None, 16, 16, 16)    3088        Normalization[0][0]
____________________________________________________________________________________________________
conv2 (Convolution2D)            (None, 8, 8, 32)      12832       conv1[0][0]
____________________________________________________________________________________________________
conv3 (Convolution2D)            (None, 4, 4, 64)      51264       conv2[0][0]
____________________________________________________________________________________________________
flat (Flatten)                   (None, 1024)          0           conv3[0][0]
____________________________________________________________________________________________________
drop1 (Dropout)                  (None, 1024)          0           flat[0][0]
____________________________________________________________________________________________________
elu1 (ELU)                       (None, 1024)          0           drop1[0][0]
____________________________________________________________________________________________________
fully_connected1 (Dense)         (None, 512)           524800      elu1[0][0]
____________________________________________________________________________________________________
drop2 (Dropout)                  (None, 512)           0           fully_connected1[0][0]
____________________________________________________________________________________________________
elu2 (ELU)                       (None, 512)           0           drop2[0][0]
____________________________________________________________________________________________________
output (Dense)                   (None, 1)             513         elu2[0][0]
====================================================================================================
Total params: 592,497
Trainable params: 592,497
Non-trainable params: 0
```

Fig. 8 Sample images from the training datasets

Table 1 Steering angle dataset information

Dataset name	No of images	Size (GB)	Data format
D1	63,000	3.1	frame.jpg angle,year-mm-dd hr:min:sec:millisec
D2	45,500	2.2	frame.jpg angle

Table 2 RMSE values for models

Model name	Dataset	Training	Validation	Test
Nvidia DAVE-2	D1	0.0629	0.0989	0.0981
	D2	0.0743	0.0997	0.0986
Comma.ai	D1	0.0789	0.1120	0.1131
	D2	0.0792	0.1217	0.1372

this is done so that the image that we are passing does not contain any distortions. The parameters which are to be identified can be internal such as the focal length, or it can be external such as the amount by which it is rotated or moved across (Fig. 9).

There are two sets of coordinates which we deal with when it comes to camera calibration: one is the set of coordinates which is observed by the world (with respect to some reference point) called as world coordinates and the other is set of coordinates with respect to the camera called as camera coordinates. Let us represent the world coordinates by (X_C, Y_C, Z_C) and the camera coordinates by $X_C, Y_C, Z_C(X_W, Y_W, Z_W)$. The camera coordinates are related to the world

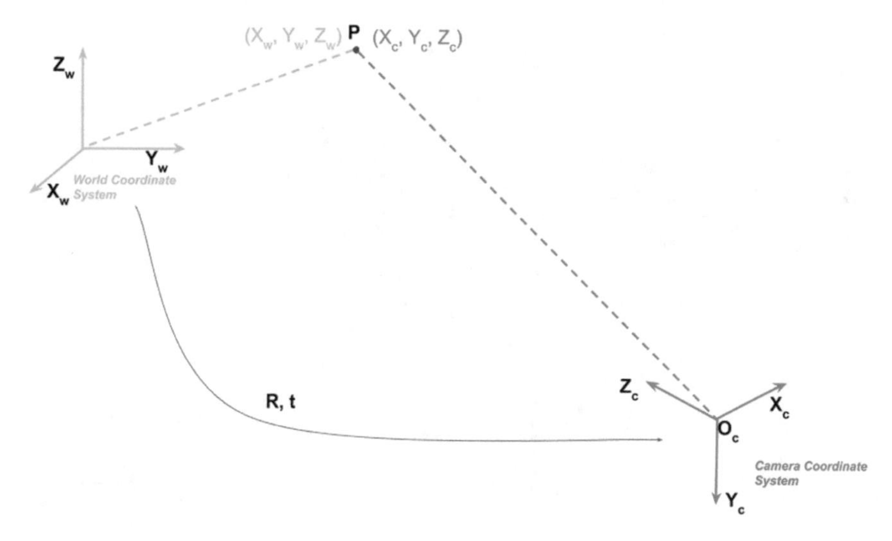

Fig. 9 Representation of a point with respect to world coordinates and camera coordinates

coordinates by the following equation:

$$\begin{bmatrix} X_C \\ Y_C \\ Z_C \end{bmatrix} = R \begin{bmatrix} X_W \\ Y_W \\ Z_W \end{bmatrix} + t \tag{1}$$

where R is the factor by which it has been rotated and t is the factor by which it has been translated. Converting this to a shorthand form with the use of homogeneous coordinates we get the following:

$$\begin{bmatrix} X_C \\ Y_C \\ Z_C \end{bmatrix} = [R|t] \begin{bmatrix} X_W \\ Y_W \\ Z_W \\ 1 \end{bmatrix} \tag{2}$$

where we define $[R|t]$ as the extrinsic matrix P. The homogeneous representation is extremely helpful when we are referring to points which are present far way i.e. points at infinity can be represented as (1, 1, 1, 0) where the coordinate transformation would be (1/0, 1/0, 1/0).

Once the camera coordinates are established, we now need to convert the representation from its three-dimensional form (X_C, Y_C, Z_C) to that of two-dimensional image place coordinate (x, y) as shown in Fig. 10. O_c represents the center of the camera and the image plane is formed at the focal length f. By the property of similar triangles, we can establish that $x = f * \frac{X_C}{Z_C}$ and $y = f * \frac{Y_C}{Z_C}$. In matrix form, this can be converted into the following:

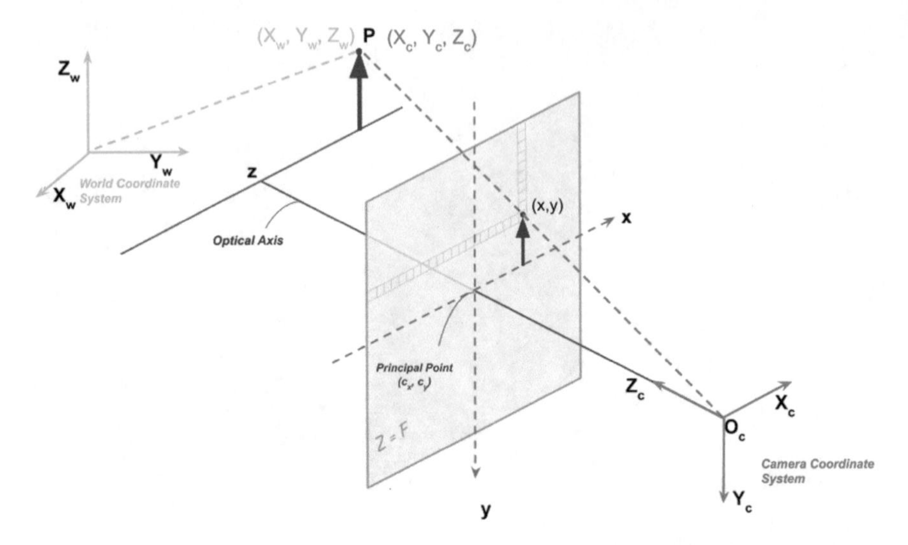

Fig. 10 Relation between camera coordinate system and image plane system

$$\begin{bmatrix} x' \\ y' \\ z' \end{bmatrix} = \begin{bmatrix} f & 0 & 0 \\ 0 & f & 0 \\ 0 & 0 & 1 \end{bmatrix} \begin{bmatrix} X_C \\ Y_C \\ Z_C \end{bmatrix} \tag{3}$$

However, this does not account for the varying focal lengths (f_x, f_y) that can exist along the x and the y axis and the center of the camera may not be in line with the center of the image plane (c_x, c_y). Also accounting for the skew s, we get the following:

$$\begin{bmatrix} x' \\ y' \\ z' \end{bmatrix} = \begin{bmatrix} f_x & s & c_x \\ 0 & f_y & c_y \\ 0 & 0 & 1 \end{bmatrix} \begin{bmatrix} X_C \\ Y_C \\ Z_C \end{bmatrix} \tag{4}$$

In shorthand this can be written as:

$$\begin{bmatrix} x' \\ y' \\ z' \end{bmatrix} = I \begin{bmatrix} X_C \\ Y_C \\ Z_C \end{bmatrix} \tag{5}$$

where I is the intrinsic matrix. With the application of homogeneous coordinates, we get $x = \frac{x'}{z'}$ and $y = \frac{y'}{z'}$. Thus, in entirety including the extrinsic matrix R, the intrinsic matrix I and the amount by which is moved t, this process can be written as:

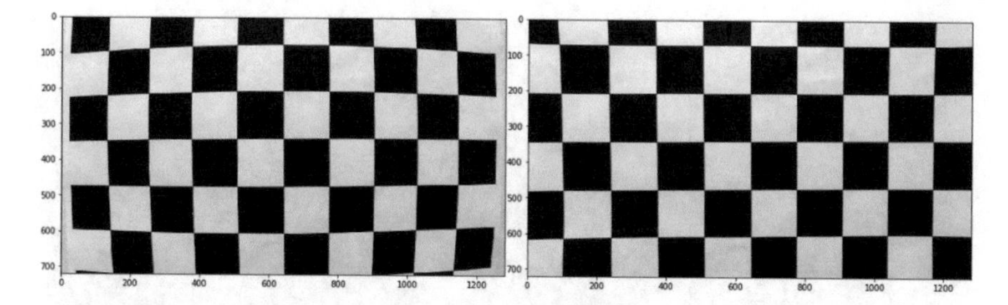

Fig. 11 Distorted and undistorted versions of the chessboard image

$$\begin{bmatrix} x' \\ y' \\ z' \end{bmatrix} = I[R|t] \begin{bmatrix} X_W \\ Y_W \\ Z_W \\ 1 \end{bmatrix} \tag{6}$$

It is now clear that the calculation of extrinsic and the intrinsic parameters is imperative to undistort the image. Using a chess board to remove calibration is one of the well-known methods to calculate these parameters. This is because the points are very well defined in a chess board pattern. First, we establish the points which we want to focus on using the world coordinates present in the chessboard. Then we use different angles and capture the same image. OpenCv exposes a first-class method called as findChessboardCorners [13] which gives the image plane coordinates x and y. We pass these coordinates to another function calibrate Camera [13] which gives us the parameters needed to calibrate the camera. Since these parameters are now identified we can use them to generate the undistorted image (Fig. 11).

Since the camera is calibrated, we can now use the same to them on road images taken by the camera (Fig. 12).

Perspective transform. Perspective transform is needed to find the top view of the road [14]. This can be obtained if we can map the coordinates of the corners of the trapezium, which is our region of interest, into coordinates of a rectangle such that the top view is shown (Fig. 13).

Mathematically, this can be represented as:

Fig. 12 Distorted and undistorted version of road image

Fig. 13 Region of interest on road

$$\begin{bmatrix} x' \\ y' \\ 1 \end{bmatrix} = sH \begin{bmatrix} x \\ y \\ 1 \end{bmatrix} \tag{7}$$

where (x', y') are the new coordinates, (x, y) are the old coordinates, H represents a homographic matrix and s is a scalar component. This transformation is also done with the help of a first class OpenCv function getPerspectiveTransform [14].

The images are then applied of this transformation using warpPerspective [14] (Fig. 14).

Mapping pixels to metres. The lane lines were approximately 12 feet apart. We approach this problem in x and y dimensions separately. The images are transformed to the HLS color space, and the lane markings are identified using a threshold color component specifying that yellow color space conforms to a lane marking. For the x dimension, the centroids of the lane markings are calculated using the concept of moments and the distance between them is noted. Now that since we know that they are 12 feet apart, we use the distance as a ratio to map this. For y we use the homographic matrix defined earlier to obtain the conversion. This can be generalized to any given road image, given that we have the reference point in lane markings identifying the separation in distance metric (Fig. 15).

Let us summarize the above elaborated techniques to apply the same on an image or video pipeline to identify roads:

- First the original image is subjected to undistortion by the parameters retrieved by the camera calibration procedure.
- Post that, gaussian blur, a type of smoothening using gaussian function is used to filter out the higher frequencies.

Fig. 14 Top view warped image of the region of interest

- Then in order to uncover the lane edge markings, we use color transform and sobel filtering [15] and get the binary counterpart. The binary counterpart has only edge components of the image, which we are interested in.
- Now we convert the region of interest into a top down view using perspective transformation.
- Then we establish a mask for filtering out the non-road marking pixels using a sliding window approach.
- Thereafter the mask is applied to the binary image to clearly identify the lane markings.

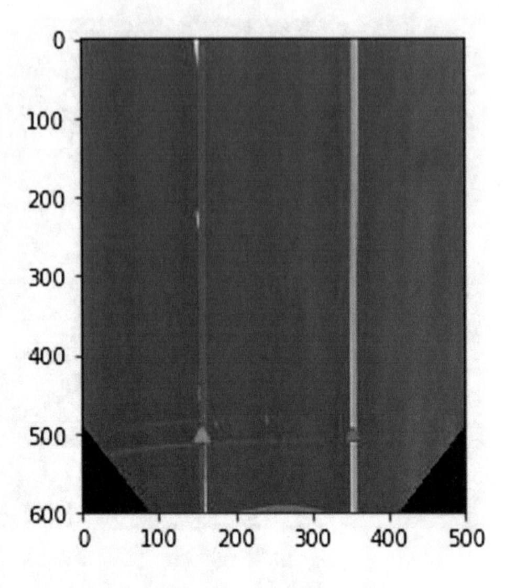

Fig. 15 Identification of lane markings and their centroids

- Now that the lane markings are identified, we subject the same to fitting of quadratic lines.
- We color the identified line markings and area in between and unwarp the image.
- The pixels are mapped and the distances between the center of the lane and other required objects are calculated. Once the distances are mapped, we can analyze the pixel shifts and calculate actual velocities (Fig. 16).

The mentioned steps in the image pipeline are applied in order to identify road markings in this chapter. The advantage of identifying the road markings is that it could verify if the steering predicted by the vehicle is correct or not. By having this as an auxiliary function to the underlying deep learning network, we can largely eliminate wrong steering actions predicted.

3.3 Object Detection Using YOLO-V3

Object detection using neural networks is an interesting problem because objects may have different sizes in the images under consideration. Using deep learning methods, this task largely falls under one of R-CNN (Recurrent convolutional neural networks) [16], SSD (Single shot detection) [19] or YOLO (You only look once) [20]. R-CNN as proposed by Girshick et al. [16] follows a two-stage approach, the first stage consists of identifying the bounding boxes where objects can be found, and the second stage is to pass the identified region of interests for classification. In the first step the image is broken down into approximately 2000 region proposals and each of these proposals are fed into a CNN for classification. This distinction as two steps is an arduous process which takes a lot of time. Girshick et al. [17] introduced

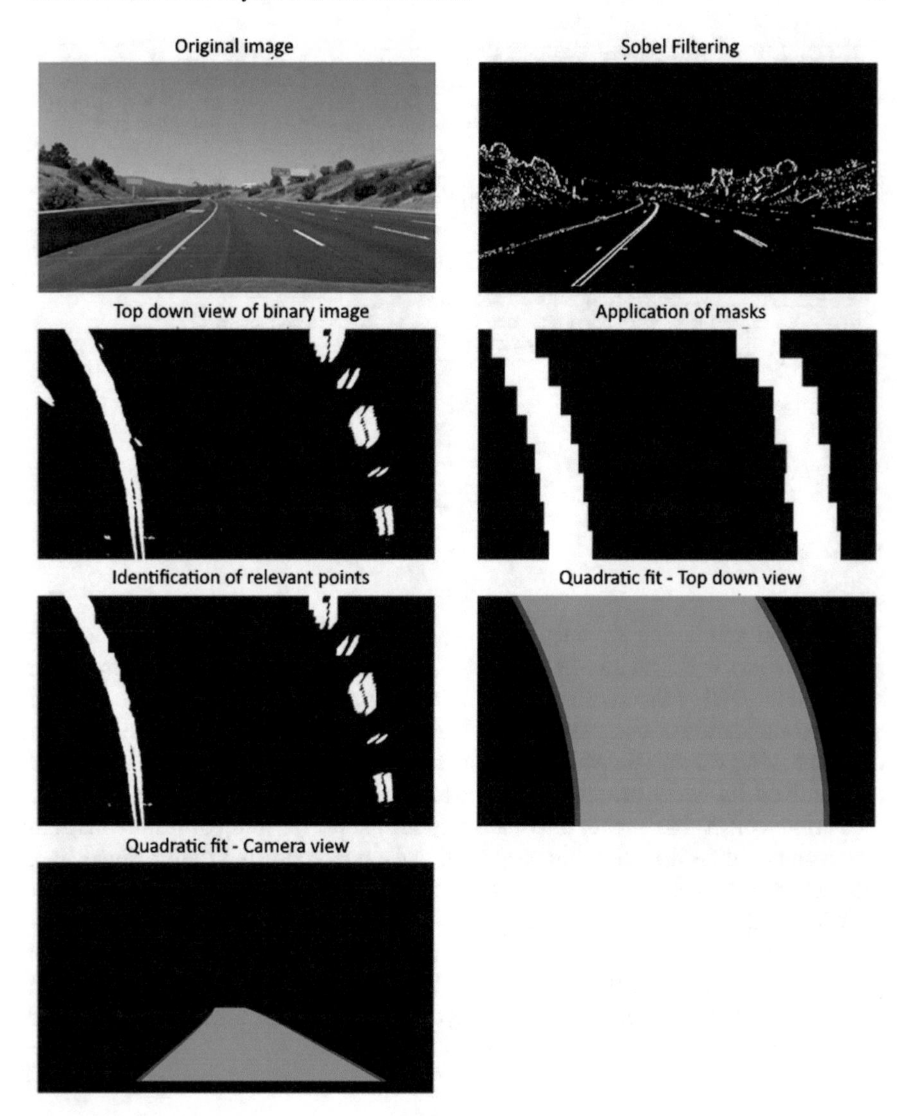

Fig. 16 Image pipeline tasks

another detector called as Fast R-CNN in which instead of the splitting the image into region proposals, the image is converted into a feature map and then the proposal regions are looked upon. This was significantly faster than its predecessor. However, these algorithms still had a dependency on selective search which did not make them end-to-end. Girshick et al. [18] solved this problem in Faster R-CNN by using a region proposal network over a selective search algorithm (Fig. 17).

To increase the speed, the intuition would be to convert the double stage process into a single stage process, and this is what SSD [19] and YOLO [20] achieves.

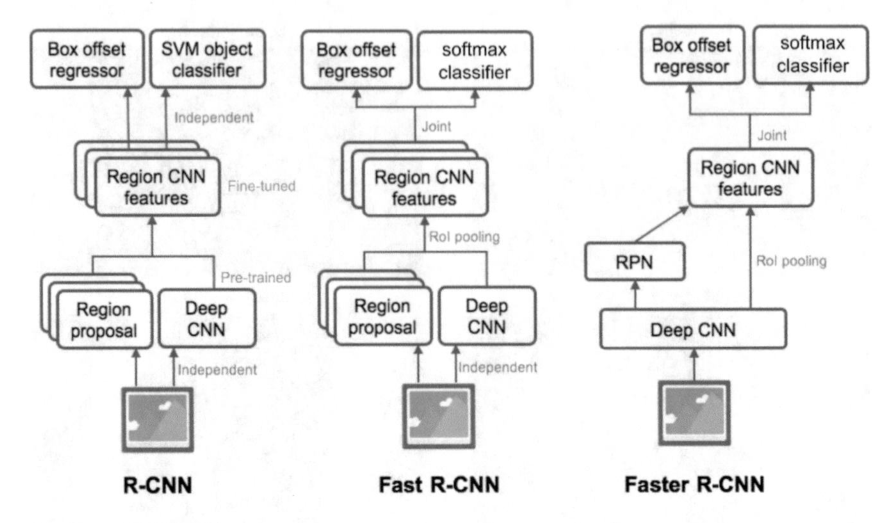

Fig. 17 R-CNN architectures

The tradeoff here is the accuracy of the system. SSD networks are broken up into a backbone network and an SSD head. SSD networks use a pyramidal approach in which all the CONV layers are connected to the output layer. This solves the issue of varying boundaries of bounding boxes. In each layer, a detection is made aimed at objects having varying dimensions. There are predefined anchor boxes, also called as default boxes, which are present having a definite size and position with respect to the current cell. These anchor boxes are responsible for the tiling of feature maps in a convolutional fashion. Since at each level, we are targeting a different dimension, the anchor boxes are rescaled to match the receptive field size (Fig. 18).

YOLO [20] is one of most popular networks considered for object detection. The backbone network is inspired from GoogLeNet in which the inception module is altered by CONV layers. It has a total of 24 convolutional layers which perform feature extraction and 2 dense layer which performs predictions. The base architecture is called the DarkNet architecture (Fig. 19).

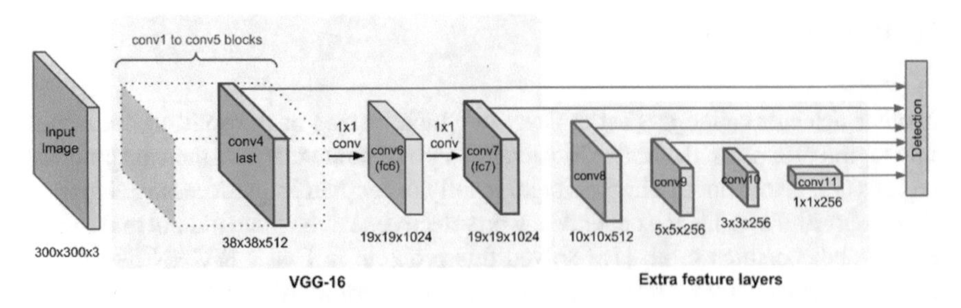

Fig. 18 SSD architecture with VGG-16 backbone network

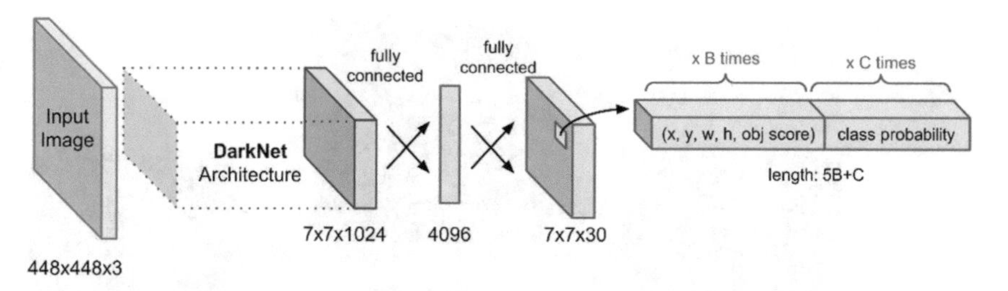

Fig. 19 YOLO architecture

A cell is said to be held responsible for detecting an object if that object's center falls in the cell. There are three things which are predicted by the cell:

- Bounding box coordinates: The bounding box coordinates is given by a set of 4 values—x coordinate of the center, y coordinate of the center, height of the bounding box and the width of the bounding box. These values are normalized such that the values fall between (0, 1].
- Confidence score: The confidence score which gives the measure of how likely an object exists in the cell is given by the product of the probability of finding an object and the Intersection over Union (IOU) which is a measure of how much the detected region aligns with the ground truth upon the total area spanned by them.
- Probability of associated class: Probability of the object being associated with one of the class members with which we trained with. Mathematically this is $Pr(ObjectbelongingtoclassX_i|Objectexists)$ where X represents the class and i ranges from 1 to k total number of classes.

Let us consider an image which is split into $C * C$ cells. If there are B bounding boxes, then there are 4 values which correspond to the bounding box coordinates, 1 confidence score and k possible conditional probabilities, which means there are a total of $C * C * (5B + K)$ possibilities, which is the size of the output dimension vector (Fig. 20).

In this chapter we used the YOLO-V3 [21] which is based off the darknet-53 architecture by stacking up additional 53 layers. This was trained on the COCO [22] dataset consisting of 80 labels. The labels of interest in this problem are person, bicycle, car, motorbike, bus, train, truck, traffic light, stop sign, parking meter, bench, cat, dog and cow (Fig. 21).

3.4 Combined Pipeline Output

The three modules discussed are combined to form a single system which gives the following:

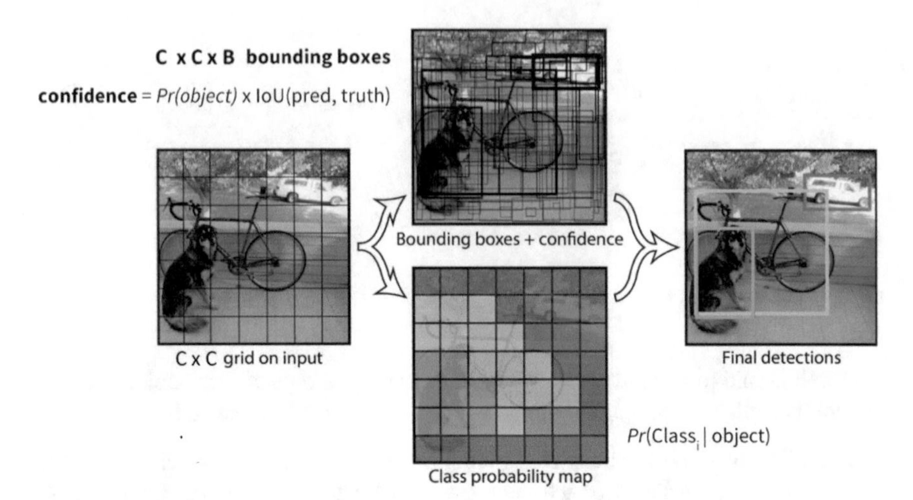

Fig. 20 YOLO output tensor workflow

Fig. 21 Application of YOLO-V3 to detect interested objects

- Steering angle—The NVIDIA end-to-end CNN model is used to calculate the steering angle.
- Lane curvature—The road detection module helps in calculating the lane curvature.
- Offset from lane center—The road detection module helps in calculating the deviation from lane center.
- Object identification in vicinity along with velocities—The object detection module helps in identifying objects, and once identified the pixel movement is used to calculate the relative velocities and thereafter final velocities.
- Acceleration safety—Assesses current lane and relative velocities in current lane to suggest whether the vehicle can be accelerated or not (Fig. 22).

Fig. 22 Combined pipeline output

The system is now able to comprehend much more information about the surroundings than a blind end-to-end system. The output is the steering angle, radius of curvature of the road with lane deviation, and objects present in the view with their velocities. It can also identify the erroneous steering angles which are predicted part of the system. Thus, by having a foundation of an end-to-end model and by having auxiliary functions which assist them, the system has ameliorated.

4 Conclusion and Future Work

In this chapter we proposed a system which was able to predict the steering angle, analyze road markings and suggest radius of curvature of the road with lane deviation, and identify objects present in the view with their velocities. Rather than having a dependency on one single architecture to handle this complex task, having auxiliary modules in addition to a base CNN architecture can enable the autonomous vehicle to interpret more information and act accordingly. Future work around this can be to improve the base model and the latency of the pipeline by parallelizing the tasks since they are independent of each other. Although the current training for the models did involve an amount of data augmentation, it is nowhere close to replicating all given scenarios. It is thus imperative to create fool-proof systems and thus the training data set must be continuously updated by new encounters.

References

1. Levinson J, Askeland J, Becker J, et al (2011) Towards fully autonomous driving: systems and algorithms. In: 2011 IEEE intelligent vehicles symposium (IV). https://doi.org/10.1109/ivs.2011.5940562
2. Krok A (2020) Tesla trails Waymo, Cruise and others in self-driving strategy, study claims. In: Roadshow. https://www.cnet.com/roadshow/news/self-driving-study-navigant-research-tesla-waymo-cruise/. Accessed 28 Apr 2020
3. Flämig H (2016) Autonomous vehicles and autonomous driving in freight transport. In: Autonomous driving, pp 365–385. https://doi.org/10.1007/978-3-662-48847-8_18
4. Autonomous Driving. In: Daimler. https://www.daimler.com/innovation/product-innovation/autonomous-driving/. Accessed 18 May 2020
5. J3016B: Taxonomy and Definitions for Terms Related to Driving Automation Systems for On-Road Motor Vehicles—SAE International. In: J3016B: taxonomy and definitions for terms related to driving automation systems for on-road motor vehicles—SAE international. https://www.sae.org/standards/content/j3016_201806/. Accessed 28 Apr 2020
6. Liu S, Tang J, Zhang Z, Gaudiot J-L (2017) Computer architectures for autonomous driving. Computer 50:18–25. https://doi.org/10.1109/mc.2017.3001256
7. Cohen J (2019) Self-driving cars—the guide. In: Medium. https://towardsdatascience.com/self-driving-cars-the-guide-f1f427b9656b. Accessed 18 May 2020
8. Pomerleau DA (1989) ALVINN: an autonomous land vehicle in a neural network. In: Touretzky DS (ed) Advances in neural information processing systems, vol 1., Morgan Kaufmann
9. DAVE: autonomous off-road vehicle control using end-to-end learning. In: CBLL, research projects, computational and biological learning lab, Courant Institute, NYU. https://cs.nyu.edu/~yann/research/dave/. Accessed 28 Apr 2020
10. Bojarski M, Firner B, Flepp B, et al (2016) End-to-end deep learning for self-driving cars. In: NVIDIA developer blog. https://devblogs.nvidia.com/deep-learning-self-driving-cars/. Accessed 28 Apr 2020
11. Comma.ai (2016) comma.ai/research. In: GitHub. https://github.com/commaai/research. Accessed 10 May 2020
12. SullyChen (2019) SullyChen/driving-datasets. In: GitHub. https://github.com/SullyChen/driving-datasets/. Accessed 10 May 2020
13. Camera Calibration and 3D Reconstruction. In: Camera calibration and 3D reconstruction—OpenCV 2.4.13.7 documentation. https://docs.opencv.org/2.4/modules/calib3d/doc/camera_calibration_and_3d_reconstruction.html. Accessed 10 May 2020
14. Geometric Image Transformations. In: Geometric image transformations—OpenCV 2.4.13.7 documentation. https://docs.opencv.org/2.4/modules/imgproc/doc/geometric_transformations.html. Accessed 10 May 2020
15. Kanopoulos N, Vasanthavada N, Baker R (1988) Design of an image edge detection filter using the Sobel operator. IEEE J Solid-State Circuits 23:358–367. https://doi.org/10.1109/4.996
16. Girshick R, Donahue J, Darrell T, Malik J (2014) Rich feature hierarchies for accurate object detection and semantic segmentation. In: 2014 IEEE conference on computer vision and pattern recognition. https://doi.org/10.1109/cvpr.2014.81
17. Girshick R (2015) Fast R-CNN. In: 2015 IEEE international conference on computer vision (ICCV). https://doi.org/10.1109/iccv.2015.169
18. Ren S, He K, Girshick R, Sun J (2017) Faster R-CNN: towards real-time object detection with region proposal networks. IEEE Trans Pattern Anal Mach Intell 39:1137–1149. https://doi.org/10.1109/tpami.2016.2577031
19. Liu W, Anguelov D, Erhan D, et al (2016) SSD: Single shot MultiBox detector. In: Computer vision—ECCV 2016 lecture notes in computer science, pp 21–37. https://doi.org/10.1007/978-3-319-46448-0_2
20. Redmon J, Divvala S, Girshick R, Farhadi A (2016) You only look once: unified, real-time object detection. In: 2016 IEEE conference on computer vision and pattern recognition (CVPR). https://doi.org/10.1109/cvpr.2016.91

21. Redmon J, Farhadi A (2018) YOLOv3: an incremental improvement. In: Computer vision and pattern recognition (cs.CV). arXiv:1804.02767
22. Common Objects in Context. In: COCO. http://cocodataset.org/#home. Accessed 18 May 2020

Deep Learning Technologies to Mitigate Deer-Vehicle Collisions

Md. Jawad Siddique and Khaled R. Ahmed

Abstract Deer-Vehicle Collisions (DVCs) are a growing problem across the world. DVCs result in severe injuries to humans and result in loss of human lives, properties, and deer lives. Several strategies have been employed to mitigate DVCs and include fences, underpasses and overpasses, animal detection systems (ADS), vegetation management, population reduction, and warning signs. The main aim of this chapter is to mitigate deer-vehicle collisions. It proposes an intelligent deer detection system using computer vision and deep learning techniques. It warns the driver to avoid collision with deer. The generated deer detection model achieves 99.3% mean average precision (mAP@0.5) and 78.4% mAP@0.95 at 30 frames per second on the test dataset.

Keywords Deep learning · Deer-vehicle collisions · YOLOv5 model

1 Introduction

Deer-Vehicle Collisions (DVCs) are a global problem that is resulting in serious injuries to humans and results in the loss of human and deer lives. Deer are more active and less attentive during the mating and hunting seasons. Roadside deer activity such as feeding and strolling along the roadside has a significant correlation with DVCs. According to the Federal Highway Administration, currently, a 4-million-mile system of public roads in the United States is used by more than 227 million vehicles every year in 2018 [1]. From 2010–2050, it is estimated that the world will need to add nearly 25 million paved road lane kms (approximately a 60% increase [2] that is).

Equivalent to encircling the earth by about 600 times [3]. Global travel in 2050 will double that of 2010 travel levels; passenger travel will account for 70% of this growth, and about 90% of travel growth is expected in countries such as the USA.

Md. Jawad Siddique · K. R. Ahmed (✉)
Southern Illinois University, Carbondale, IL, USA
e-mail: khaled.ahmed@siu.edu

K. R. Ahmed et al. (eds.), *Deep Learning and Big Data for Intelligent Transportation*, Studies in Computational Intelligence 945,
https://doi.org/10.1007/978-3-030-65661-4_5

DVCs are expected to increase with time as mammal-vehicle collisions have increased significantly since the 1970s [4]. The frequency of DVCs escalates throughout Europe. Each year in Germany, about 200,000 roe deer collide with vehicles, which is almost 20% of the hunters' roe deer harvest. These collisions led to approximately 3000 injured people, with 50 fatalities and costs of about 490 million € [5]. In Slovenia, between 4000 and 6000 roe deer were killed on roads each year [6]. In the USA, the National Highway Traffic Safety Administration (NHTSA) estimates that there are about one million DVCs annually [7], killing about 200 people [8] and cost around $1 billion a year in revenue [9]. In 2018, these deaths occurred most often during July-September [10] as shown in Fig. 1. According to the State Farm report of 2019 [11], the top ten states of USA where deer collisions are most likely are West Virginia (1 in 38), Montana (1 in 48), Pennsylvania (1 in 52), Wisconsin (1 in 57), Iowa (1 in 55), South Dakota (1 in 54), Minnesota (1 in 64), Michigan (1 in 60), Wyoming (1 in 56) and Mississippi (1 in 61). Over the past 10 years, the state of Texas had the highest numbers of deaths from collisions with animals. In 2018, there were 319,146 crashes involving motor vehicles in Illinois. Crashes involving deer accounted for 4.9% which is 15,635 crashes of overall crashes in 2018. The number of fatal collisions has doubled in 2018, reaching 630 injuries and 87 incapacitating injuries (A-Injury) (Illinois 2018) [12].

Several strategies have been used to mitigate DVCs and include vegetation management, fences, underpasses and overpasses, population reduction, warning signs and animal detection systems (ADS). These strategies vary in their efficacy. These strategies may help to reduce DVCs however, they are not always easily feasible due to false alarm, high cost, unsuitable terrain, land ownership, and other factors. DVCs are increasing due to the increase in number of vehicles and the absence of intelligent highway safety and alert systems. This chapter introduces an automated DVCs mitigation system that combines computer vision, artificial intelligent methods with deep learning techniques. It developed a deer detection model based on one-stage deep learning algorithm (Yolov5).

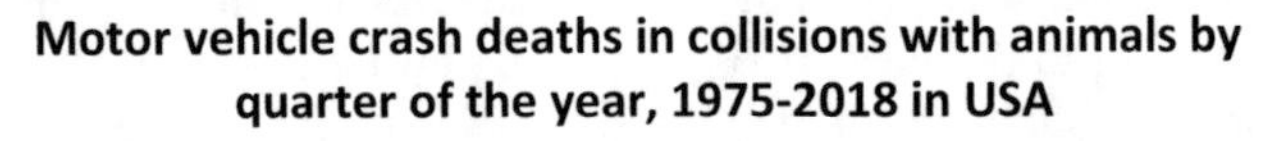

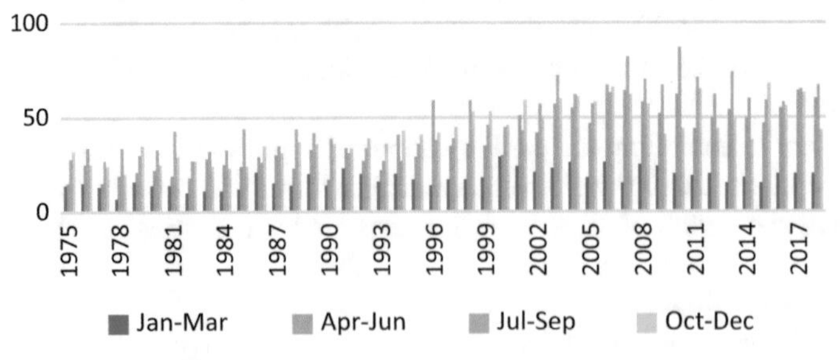

Fig. 1 Death in vehicle-animal collisions 1975–2018

2 Related Works

2.1 Traditional DVCs Mitigation Approaches

Several approaches have been implemented to handle Animals Vehicle Collision (AVC) problems. Khaled R., et al. classifies the approaches used in the past to mitigate AVC into three categories *animal-based*, *roadway-based*, and *vehicle-based* techniques [13].

The first category, roadway-based technique developed for and dedicated to the roads and highways. It includes roadway-based traditional techniques and Roadway-based Detection systems. The traditional roadway-based techniques include vegetation management, roadway fencing, underpasses, overpasses, roadway reflectors, warning signs, and highway lightening. Finder et al. recommended to remove woody vegetation along roads and highways to a width of at least 40.0 m [14]. Ward et al. reported that a two-meter-high big-game fence is effective in reducing DVC [15]. The combination of fencing and crossing structures has been shown to reduce road-kill of large animals such as deer by 83% [16, 17]. Wildlife passage structures may reduce DVCs however, they are not always easily feasible due to high cost, unsuitable territory, land ownership, and other issues. Common approaches to reducing AVC are warning signs [18]. Lighted, animated deer-crossing warning signs help to reduce drivers' speeds, however after the drivers got used to the animated signs, they lost their effect in reducing the drivers' speeds [19].

Roadway-based detection systems are designed to warn drivers/animals once animals/drivers were detected. Animals detect technologies detect large animals as they approach the road. When an animal is detected, signs are activated to warn drivers that large animal may be on or near the road at that time. Infra-red, microwave reader, laser sensors were used to detect animals. Infra-red sensors produced false detections due to strong winds and warm engines of passing vehicles. Microwave reader sensor produced false detection in spring when the snow melted and the water warmed on the pavement, spray from passing vehicles triggered warning. Laser sensors may produce false detection when anything broke the laser beam triggered the warning, including birds, dogs, and mail trucks. Even the sun could trigger the beacons depending on the time of year as sunrise and sunset angles changed [20]. A vehicle detection system consists of a small cabinet with electronics, sensors for vehicle detection, and an animal warning device. They are powered by solar panels and batteries. Once vehicles are detected, units in the roadside are activated and alert deer through a variety of noise and light signals [21]. These signals may not scare the animals as animals are able to adapt to disturbance if this is not accompanied by an immediate and real threat. Additionally, such system is not suitable for high traffic flows as the animal warnings would be running continuously. Huijser designed an Animal-detection systems (ADS) to warn drivers using flashing signs when animals are adjacent to or within a roadway, and various methods (e.g. thermal video camera and Doppler radar) have been used for detecting animals and activating signs [22]. Trina Rytwinski reported that ADS systems led to a reduction of 57% in roadkill of large mammals, and only a 1% for wildlife reflectors [16, 17]. However, ADS may suffer from false negative as animal leaving the detection area triggers warning to motorists. The false negative occurs when there is animal, but the warning system is not activated. Whereas the false positive occurs when the warning system is activated even if there is no animal. Huijser et al. stressed that ADS should be considered experimental because there is no single reliable system available for universal application [17].

The second category, animal-based includes the technologies that were installed in animals to alleviate the AVCs. Different types of collars fasten with the animal to trigger a warning system such as blinking signals. They are classified as reflective collars, radio collars, and GPS collars. Radio collar fasten only to lead animals [23] and receivers placed along the road scan for the frequencies of the individual radio collars 24 h per day. When a radio-collared animal come within about 400 m of the road, the receivers that pick up the signal activate the flashing beacons. The animals without a radio collar are only detected if radio-collared animals accompany them. Therefore, the system only works well for highly sociable species. The radio-collar system requires re-collaring effort. The batteries of the radio collars usually run out after several years, and then they must be replaced. GPS collars used to monitor large animal movements for the sake of recommending the placements of wildlife-engineered crossing structures on highways [24–26]. Khaled R., et al. developed Camel-Vehicle Accident Avoidance System (CVAAS) in KSA [13]. They fasten solar-powered GPS collars to domestic camels to detect the camel position, direction, and movement. The collar transmits the camel position to the CVAAS server using General Packet Radio System (GPRS). Once a camel enters the dangerous zone around the highway, the CVAAS system broadcasts a warning to drivers in the affected area. Deploying and maintaining the CVAAS system in large scale are not easily feasible due to high cost.

The third category, vehicle-based includes the technologies that are equipped into vehicles to reduce the AVC. They generally classified into two major groups: warning whistles (e.g. air-activated whistles) and infrared detection systems. The warning whistles attached to vehicles produce ultrasonic frequencies to warn animals of approaching vehicles. University of Georgia researchers reported that warning whistles produce improper function and sufficient sound intensity to be audible to deer in roadway conditions [27]. The infrared detectors warn drivers when a large animal is detected within a certain range from the sensors attached to the vehicle [28]. This range should be sufficient to enable for the driver to stop his/her vehicle before colliding with the detected animal [29].

Therefore, there is a need for not only lowering of the expense of DVCs mitigation systems but also to propose and develop new approaches, including approaches that would increase the level of automated electronic observance and awareness for the drivers. This chapter introduces an automated DVCs mitigation system that combines computer vision, artificial intelligent methods with deep learning techniques.

2.2 *Machine Learning Deer Detection*

Several researches have been proposed to detect wildlife animals on or nearby roads using several technologies such as movement and infrared sensors, surveillance video cameras and Global Positioning Systems (GPS) [13]. The present technology that can help to mitigate deer vehicle collision is machine learning especially deep learning techniques. Some ML and computer vision techniques were used to recognize and detect objects such as wildlife animals with a focus to count, locate and classify wildlife animals accurately [30, 31]. This area of research used computer vision algorithms to extract features from images/videos and then use supervised/unsupervised machine learning algorithms to predict the labels of a given images. Videos/images may include animals having similarities in appearance with background information, thus the use of classical computer vision algorithms involve tedious feature engineering [32]. In addition, several studies have been conducted to extract moving objects from images, which captured with a fixed camera, using logical operations on positional differences [33]. Creating an effective model to capture the complex background motion and texture dynamics and segment the foreground animals is the key challenge [34]. Thus, accurate and roust images/vide object detection and segmentation in dynamic scene and under different illumination and weather conditions are challenges. Deep learning (DL) is an alternative to resolve the challenges that some ML approaches face, the performance of object detection has been improved greatly with the application of DL techniques. DL is useful since we do not need to explore various feature extractors. It allows computational models consisting of multiple hierarchical layers to learn complex problem such as deer detection. Backbone networks are used as a basic feature extractor which can be chosen according to the requirements such as efficiency or accuracy [35]. DL object detectors are classified into two categories: Two-stage detectors, such as Faster R-CNN [36]. In first stage, they generate region proposals network (RPN) that proposes candidate object bounding boxes. In second stage, features are extracted by Ripoll (RoI Pooling) operation from each candidate box for the following classification and bounding-box regression tasks. The second is one-stage detector, such as YOLO (you only look once) [37] and SSD [38]. The one-stage detectors propose predicted boxes from input images directly without region proposal step. Thus, two-stage detectors achieve high object recognition accuracy and localization, whereas the one-stage detectors attain high inference speed [35]. Backbone networks extract features of input images and output feature maps of the corresponding input image. Most of the backbone network layers are for detection task while the last fully connected layers are for classification task. Deeper backbone networks capture rich features and improve the accuracy of object detection tasks. However, lightweight backbones fit with systems that use mobile devices to detect objects. This chapter develops a deer detection model based on one-stage deep learning algorithm (Yolov5).

3 Deer Detection Methodology

Methodological approach is to determine the better object detection model for deer detection. Unified network detectors like YOLO, SSD successful in terms of speed while region-based object detectors like Faster R-CNN are its higher detection accuracies. The accuracies of one-stage are less sensitive quality of backbone networks than two-stage. Avoidance of pre-processing algorithms and using light-weight backbone networks and fewer candidate generator regions from specific convolutional layers made one-stage fast and less computational time in processing the image for object detection. On the other hand, two-stage detectors have separate convolutional layers for feature extraction, object proposals that are generated from region proposal networks and tuning hyper parameters for tuning and sharing the convolutional networks made them run in real time with better accuracies.

3.1 Deer Detection Using YOLO

Redmon et al. [37] described an efficacious real-time object detection algorithm YOLO ("You Only Look Once"). As FCNN (fully convolutional neural network), YOLO architecture passes the $n \times n$ image once through the FCNN and output is $m \times m$ prediction in comparison to other classification network like Fast performs detection on various region proposals and predicting multiple times for various regions in an image. "You Only Look Once" is derived from the fact that a single-stage network architecture is used to predict class probabilities and surround them with bounding boxes. This architecture is splitting the input image in $m \times m$ grid and generating two bounding boxes with class probabilities for each grid. Each grid cell may predict number of bounding boxes with different confidence scores that predicts the existence of an object in the box. The network predicts five components for each bounding box namely *x*, *y*, *w*, *h*, and confidence score. Relative to the grid cell location, the center of the box is represented by (*x*, *y*) coordinate. The (*x*, *y*) coordinates along with (*w*, *h*) box dimensions are normalized to fall between 0 and 1. The confidence score is the last component in the bounding box prediction. The presence or absence of deer can be ascertained from the confidence score. As in [37], we define confidence score as $Pr(Object) \times IOU_{pred}^{truth}$ where *Pr*(*Object*) is the probability of deer appearing in a grid cell and IOU is the interstation of union between the ground truth and the predicted boxes. If no deer exists in that box, the confidence score should be zero. The loss function penalizes both bounding box coordination error and classification error as shown in Eq. 5.

Since YOLO was first announced, few versions were released such as YOLOv2, YOLOv3, YOLOv4 and the latest one is YOLOv5. This book chapter implements a method for deer detection using YOLOv5 because it is fast, accurate and can be implemented initially in *PyTorch*. Since YOLOv5 is a single-stage object detector, it has three main parts model backbone, model neck and model head. The model backbone is mainly used to extract important features from the given input image. To make sure the network reuse features, and to reduce the number of network parameters, both YOLOv4 and YOLOv5 implement the Cross-Stage Partial (*CSP*) bottleneck based on DenseNet network. In addition, it is used to extract informative features from an input image [39]. The model neck generates the feature pyramid and helps to identify the same object with different sizes and scales. Therefore, Yolov5 able to detect close and far deer on the road. Moreover, the feature pyramid helps the model to perform well on unseen data. YOLOv5 implements the PA-NET neck for feature aggregation. The final detection part is the model Head. It generates final output vectors with class probabilities, confidence scores, and bounding boxes. The activation function in YOLOv5 are Leaky ReLU and Sigmoid activation functions. In middle/hidden layers of YOLOv5, the activation function is Leaky ReLU and in the final detection layer, it is Sigmoid activation function. SGD [40] and Adam [41] are the two-optimization functions used in YOLO v5 although SGD is the default function for training. In our model, predictions whose confidence score is lower than 0.5 are ignored. Most of the false predictions are filtered out in this way. Finally, non-maximum suppression algorithm is developed to filter objects with the same bounding box. Figure 2 shows the network structure of YOLOv5 which is divided into four parts, input, backbone, neck, and head. The input image is of a deer, which is processed using data augmentation techniques embedded in the code. After that, the

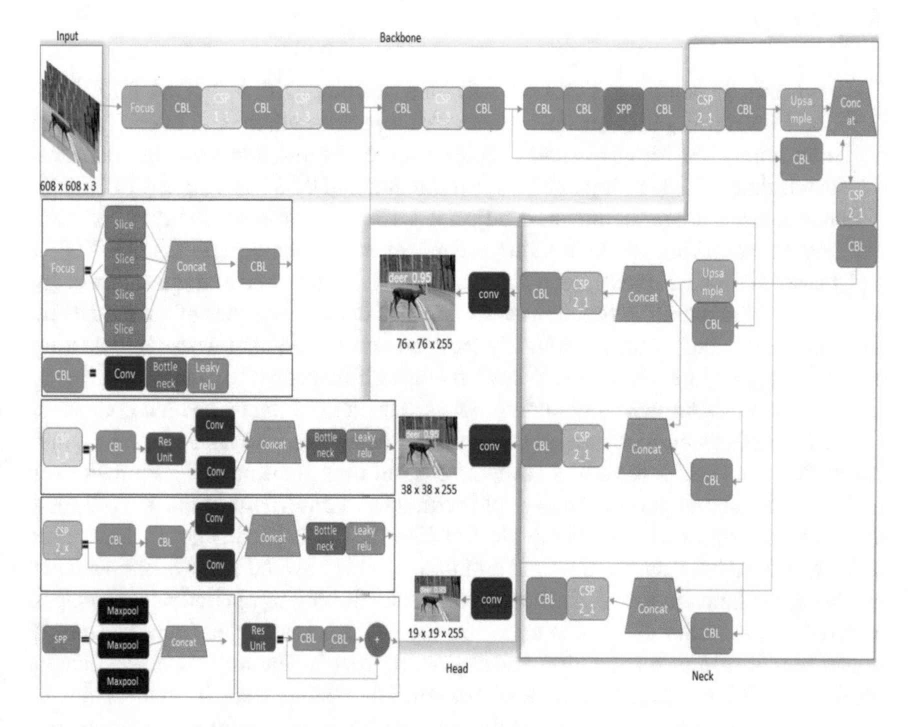

Fig. 2 Structure of YOLOv5 for deer detection

image goes to focus structure whose main key is slicing. The original *608 × 608 × 3* deer image is input into the *Focus* structure, and the slicing operation is used to first become a 304 × 304 × 12 feature map, and then after a convolution operation of 32 convolution kernels, the final output is a feature map of 304 × 304 × 32 with linear activation function. There are two cross stage partial (CSP) structure in YOLOv5 as seen in Fig. 2, CSP 1_x is applied to the backbone. To support the ability of network feature integration another CSP 2_x has been applied to the neck. Moreover, up sampling is done in the neck structure. The head includes 64 convolution kernels and results feature map 152 × 152 × 64. Finally, from Fig. 2, we see that using non-maximum suppression, we got our desired output in a bounding box predicting deer on the road.

4 Experimental Setup and Results

This section presents the experimental setup and performance evaluation of the generated models that succeed to detect deer on roads.

4.1 Setup

The environment setup for YOLOv5 was on a machine with NIVIDIA RTX 2060 GPU, and several libraries such as cudatoolkit, opencv-python, and torch. We train the Yolov5 network for about 2500 epochs on the training and validation datasets with batch size of 32, learning rate 0.01, a momentum of 0.937 and a decay of 0.0005.

Since datasets are predominant in object detection research which lays a common platform in measuring and comparing the machine learning algorithms like YOLO and Faster RCNN. Recent Deep Learning mechanisms has brought success by solving visual recognition problems and enhanced the performance of the models by data annotation and augmentation. Object detection mechanism works in detecting multiple objects in an image and label them with respective of their category in rectangular boxes known as bounding boxes. The current dataset which comprises of 1100 images categorize deer as one class for detection. The dataset used to train the models for deer detection consists of custom made and publicly sourced deer images. The custom part of the dataset is extracted from Missouri dataset [42] and images taken from Jackson County, state of Illinois, USA. The images in the dataset consist of deer both during daytime and night and it covered most of the weather condition as shown in Fig. 3. The other part of the dataset comprised of 320 images of variable sizes from *275 × 275* up to *1084 × 956* pixels sampled from the Google image search results for the term "deer crossing road". The objects in the dataset are deer of different species completely or partially visible, static, or blurred due to motion. The reason for adding the public dataset is to avoid overfitting on the custom dataset.

Fig. 3 Sample images of deer dataset

To annotate our dataset, *LabelImg,* a graphical image annotation tool is used. It saved the annotations in XMLfiles or txt files and used by ImageNet and supports XOLO format. The generated datasets are classified into train, test, and validation set through data management where 70% data in the training set, 20% in the validation set, and 10% in the testing set.

Data Augmentation

With each training batch, YOLOv5 passes training data through a data loader, which augments data online. The data loader makes three kinds of augmentations: scaling, color space adjustments, and mosaic augmentation as shown in Fig. 4. The most novel of these being mosaic data augmentation, which combines four images into

Fig. 4 Data augmentation

four tiles of random ratio that enables the model to localize deer in different size and portions in the frame.

4.2 Performance Evaluation Metrics

Several evaluation metrics such as recall, precision, accuracy, and mean average precision have been used to evaluate the performance of the generated model. The recall measures how good the model able to find all positive predictions. The precision measures how accurate the model predicts deer. The accuracy to detect deer is defined as the ratio of correctly predicted deer images by the total tested images/frames and measured as follows:

$$Recall = \frac{TP}{TP + FN} \quad (1)$$

$$Precision\ p(r) = \frac{TP}{TP + FP} \quad (2)$$

$$Accuracy\ A = \frac{TP + TN}{(TP + TN + FN + FP)} \quad (3)$$

$$AP = \int_{0}^{1} p(r)dr \quad (4)$$

$$GIoU = IoU - \frac{|C \backslash (A \cup B)|}{|C|} \quad (5)$$

where *TP* stands for True Positives, *TN* stands for True negative, *FN* stands for False Negative and *FP* stands for False Positive. The average precision (AP) is the integral over the precision p(r) as shown in Eq. 4, where precision and recalls area always between 0 and 1. The frame-based constraints are shown in Fig. 5. If the bounding box region contains the foreground object (deer), then the frame demonstrates true positive. If the object is not present inside the bounding box, then frame is considered as false positive. The frame shows false negative if target object missed by the bounding box. Finally, if object is not existing in the frame it is termed true negative. The *GIoU* metric in Eq. 5 is used to evaluate how close the prediction bounding box is to the ground truth, where Intersection over union $IoU = \frac{|A \cap B|}{|A \cup B|}$, *A*, *B* are the prediction bounding box and ground truth bounding box respectively and C is the smallest convex hull that encloses both A and B. YOLOv5 calculates a total loss function from loss functions *GIoU*, *obj*, and *class* losses. These are carefully constructed to maximize the objective of *mAP*.

Fig. 5 Illustration of frame-based constraints

4.3 Results and Discussions

In this section, we evaluate Yolov5 on deer dataset by performing testing on images from Google, videos from YouTube and real time videos taken from camera. The algorithm loaded the generated Yolo model and processed input video stream camera with the frame resolution *1920* × *1080* px at the average of 30 frames per second, which confirms that YOLO algorithm can be used for the real time video processing since it only involves a single network evaluation, unlike other classifier-based methods. Figure 6 shows some examples of the test results using the detection model. Each of the weights produced able to locate the deer in the images with high confidence score.

Fig. 6 Output with confidence score

Figure 7a–c shows the generated (Yolov5) model results when trained till 2500 epochs. It was observed that at 1500th epoch precision was 97.48%, recall was 1 and *mAP* reached to 99.5%. Intentionally, we have stopped training at 2500th to avoid overfitting. Figure 7a–d shows how the model is trained well where *mAP*@0.5 reached *99.3%* and *mAP@0.95* equal *78.4%*. Figure 7d and e shows that object loss function and generalized intersection over union loss function are decreasing for training data.

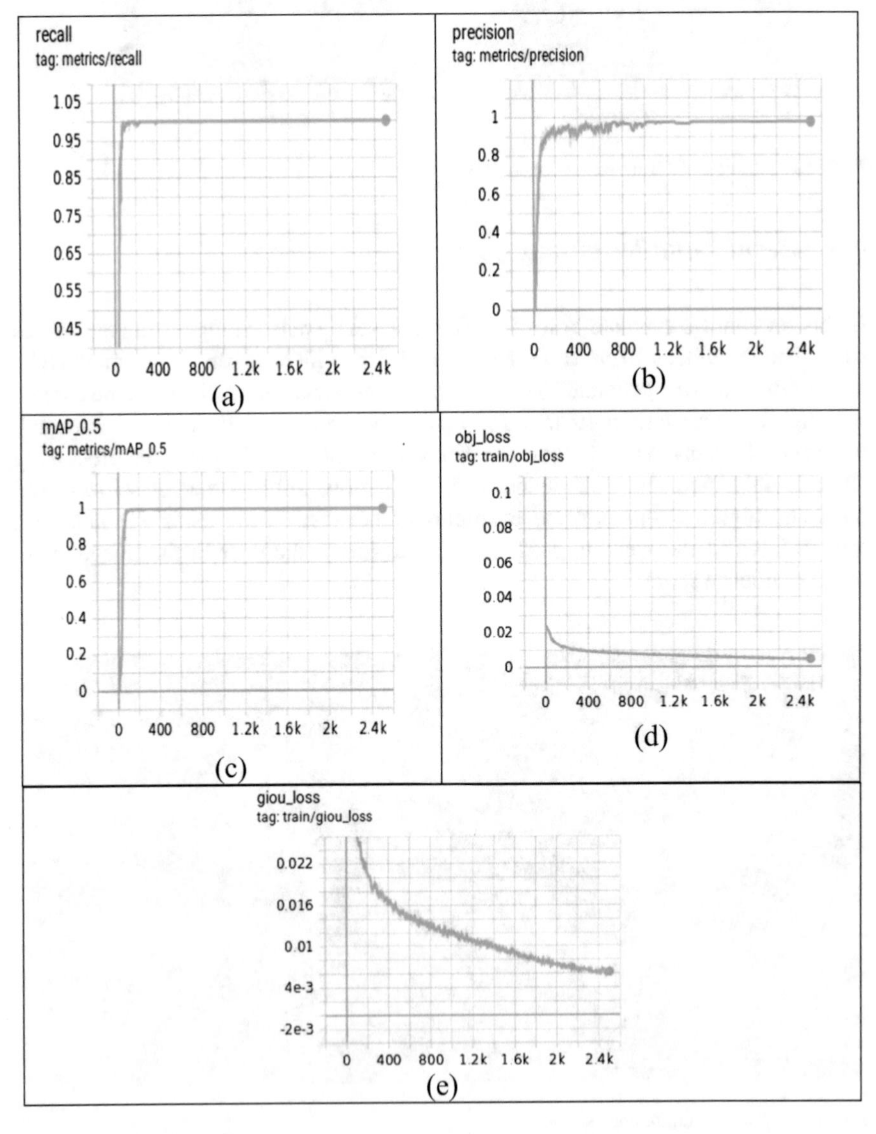

Fig. 7 Experiment results

Fig. 8 Detecting deer in herd

Deer moves as a herd, Yolov5 failed to detect all deer in the image as shown in Fig. 8. However, the generated model succeeds to detect some of deer since we set 0.5 as the minimum threshold. This failure to detect all the deer in a herd will not affect our goal to mitigate deer vehicle collision since we able to detect some of the deer in a herd.

5 Conclusion and Future Work

In this book chapter, we have presented a deep learning-based deer detector with the focus on achieving highest possible detection speeds without significantly sacrificing on detection quality. We established deer dataset that includes deer crossing roads in several different illumination and weather conditions. Data augmentation technique has been used to increase the dataset size. This chapter developed a real-time deer detector based on YOLOv5 which succeed to detect deer fast and accurately. The results show that the generated deer detection model achieves a *97.48% mAP* and *100%* recall rate. It achieves a detection speed of 30 FPS, making the real-time detection of deer more effective. In some scene when deer are far from in the video footage the generated model failed to detect it. Our future work is to adapt the Yolov5 to improve its accuracy in detection small and far deer in the scene. We will increase the size of the deer dataset through including more diverse scenarios/scenes that cover different weather and lighting conditions. In addition, we will provide comparison with other deep learning networks such as faster-RCNN with different backbones. Besides, we will develop mobile application which can detect deer from vehicle and signals the driver about the distance of the deer from the vehicle. Furthermore, we will extend our method to combine deep learning techniques to develop a unique method for detecting deer on road to mitigate deer vehicle collision.

References

1. Federal Highway Administration, Highway Statistics (FHAHS). 2011–18: U.S. (2019) Department of Transportation. Available at http://www.fhwa.dot.gov/policyinformation/statistics.cfm. 13 Nov 2019
2. Dulac J (2013) Global Land Transport Infrastructure Requirements: estimating road and railway infrastructure capacity and costs to 2050. Int Energy Agency
3. Laurance WF, Clements GR, Sloan S et al (2014) A global strategy for road building. Nature 513:229–232. https://doi.org/10.1038/nature13717
4. Hill JE, DeVault TL, Belant JL (2019) Cause-specific mortality of the world's terrestrial vertebrates. Glob Ecol Biogeogr 28:680–689. https://doi.org/10.1111/geb.12881
5. Hothorn T, Brandl R, Müller J (2012) Large-scale model-based assessment of deer-vehicle collision risk. PLoS ONE 7(2):e29510. https://doi.org/10.1371/journal.pone.0029510
6. Pokorny B (2006) Roe deer-vehicle collisions in Slovenia: situation, mitigation strategy. Vet. Arhiv **76**(Suppl.):S177–S187
7. Huijser MP, McGowen PT, Fuller J, Hardy A, Kociolek A (2008) Wildlife-vehicle collision reduction study: report to Congress. U.S. Department of Transportation, Federal Highway Administration, McLean, Virginia, USA
8. Online at http://www.car-accidents.com/pages/deer-accident-statistics.html
9. State Farm (2020) 2018–2019 animal collision likelihood. http://st8.fm/animal
10. Insurance Institute for Highway Safety (IIHS), Fatality Analysis Reporting System (FARS) (2018). https://www.iihs.org/topics/fatality-statistics
11. Deer collisions across USA. https://patch.com/us/across-america/deer-collisions-across-u-s-odds-hitting-animals
12. Illinois (2018). 2018 Illinois crash facts and statistics report, Illinois Department of Transportation, 2020
13. Khaled R et al (2013) GPS-based camel-vehicle accidents avoidance system: designing, deploying and testing. Int J Innov Comput Inf Control 9(7)
14. Finder RA, Roseberry JL, Woolf A (1999) Site and landscape conditions at white-tailed deer/vehicle collision locations in Illinois. Landsc Urban Plan 44:77–85
15. Ward AL (1982) Mule deer behavior in relation to fencing and underpasses on Interstate 80 in Wyoming, Transport, Research Record 859:8-13, National Research Council, Washington
16. Rytwinski T, Soanes K, Jaeger JAG, Fahrig L, Findlay CS, Houlahan J, van der Grif EA (2016) How effective is road mitigation at reducing road-kill? A meta-analysis. PLoS ONE 11(11)
17. Huijser MP, Mosler-Berger C, Olsson M, Strein M (2015) Wildlife warning signs and animal detection systems aimed at reducing wildlife-vehicle collisions. Handbook of road ecology. Wiley, West Sussex, UK, pp. 198–212
18. Putman RJ (1997) Deer and road traffic accidents: options for management. J Environ Manag 51:43–57
19. Pojar TM, Prosence RA, Reed DF, Woodward RH (1975) Effectiveness of alighted, animated deer crossing sign. J Wildl Manag 39:87–91
20. Washington State Department of Transportation (2010) What we are doing to reduce vehicle/wildlife
21. IRD (2002) Wildlife warning system. IRD (International Road Dynamics), Saskatoon, SK, Canada. http://www.irdinc.com/english/pdf/sys/safety/Wildlife0202.pdf
22. Huijser MP, Holland T, Blank M, Greenwood M, McGowen P, Hubbard B, Wang S (2009) Comparison of animal detection systems in a test-bed: a quantitative comparison of system reliability and experiences with operation and maintenance (Project FHWA/MT-09-002/5048). Federal Highway Administration, Helena, MT
23. Carey M (2001) Addressing wildlife mortality on highways in Washington. In: 2001 Proceedings of the international conference on ecology and transportation, pp 605–610
24. McKinney T, Smith T (2007) US93 Bighorn Sheep study: distribution and trans-highway movements of Desert Bighorn Sheep in northwestern Arizona, Final report to Arizona Department of Transportation. JPA04-032T/KR04-0104TRN

25. Dodd et al (2007) Evaluations of measures to minimize wildlife-vehicle collisions and maintain permeability across highways: Arizona Route 260, Final Report to Arizona Dept. of Transportation, JPA 01-152 JPA 04-024T
26. Gagnon et al (2009) Using global positioning system technology to determine wildlife crossing structure placement and evaluating their success in Arizona, USA. In: International conference on ecology and transportation
27. Sharon V et al (2009) Deer responses to sounds from a vehicle-mounted sound-production system. J Wildl Manag 73(7):1072–1076
28. Hirota et al (2004) Low-cost infrared imaging sensors for automotive applications, In: Valldorf J, Gessner W (eds) Advanced microsystems for automotive applications, pp 63–84
29. Forslund D, Bjarkefur J (2014) Night vision animal detection. In: IEEE intelligent vehicles symposium (IV), Dearborn, Michigan, USA
30. Wilber MJ, Scheirer WJ, Leitner P, Heflin B, Zott J, Reinke D, Delaney DK, Boult TE (2013) Animal recognition in the Mojave Desert: vision tools for field biologists. In: 2013 IEEE Workshop on applications of computer vision (WACV). IEEE, pp 206–213
31. Sengar SS, Mukhopadhyay S (2017) Moving object detection based on frame difference and W4. SIViP 11(7):1357–1364
32. Ko T, Soatto S, Estrin D (2008) Background subtraction on distributions. In: Proceedings of the 10th European conference on computer vision, pp 276–289
33. Steen KA, Villa-Henriksen A, Therkildsen OR, Green O (2012) Automatic detection of animals in mowing operations using thermal cameras. Sensors 12:7587–7597
34. Zhang Z, He Z, Cao G, Cao W (2016) Animal detection from highly cluttered natural scenes using spatiotemporal object region proposals and patch verification. IEEE Trans Multimedia 18(10):2079–2092
35. Jiao L, Zhang F, Liu F, Yang S, Li L, Feng Z, Qu R (2019) A survey of deep learning-based object detection. IEEE Access 7:128837–128868
36. Ren S, He K, Girshick R, Sun J (2017) Faster R-CNN: towards real-time object detection with region proposal networks. IEEE Trans Pattern Anal Mach Intell 39(6):1137–1149
37. Redmon J, Divvala S, Girshick R, Farhadi A (2016) You only look once: unified, real-time object detection. In: Proceedings IEEE conference on computer vision and pattern recognition, pp 779–788
38. Liu W, Anguelov D, Erhan D, Szegedy C, Reed S, Fu C-Y, Berg C (2016) SSD: single shot multibox detector. In: Leibe B, Matas J, Sebe N, Welling M (eds) Computer vision—ECCV. Springer, Cham, Switzerland, pp 21–37
39. Wang C, Mark Liao H, Wu Y, Chen P, Hsieh J, Yeh I (2020) CSPNet: a new backbone that can enhance learning capability of CNN. In: 2020 IEEE/CVF conference on computer vision and pattern recognition workshops (CVPRW), Seattle, WA, USA, 2020, pp 1571–1580. https://doi.org/10.1109/cvprw50498.2020.00203
40. Darken C, Chang J, Moody J (1992) Learning rate schedules for faster stochastic gradient search. In: Neural networks for signal processing II proceedings of the 1992 IEEE workshop, Sept, pp 1–11
41. Kingma DP, Ba JL (2015) Adam: a method for stochastic optimization. In: International conference on learning representations, pp 1–13
42. Missouri University (2016) Camera-trap dataset for wildlife species. http://videonet.ece.missouri.edu/cameratrap/videonet.ece.missouri.edu/cameratrap/

Night-to-Day Road Scene Translation Using Generative Adversarial Network with Structural Similarity Loss for Night Driving Safety

Igi Ardiyanto, Indah Soesanti, and Dwiyan Cahya Qairawan

Abstract Inadequate road infrastructures in developing countries, such as road lighting, is a real life problem affecting their transportation system. A poor road lighting hinders the ability of drivers to perceive the road conditions as well as its objects which may lead to the accident. We address a problem of synthesizing day-time appearance from the night-time road scene using loss-modified generative adversarial network. Our approach basically makes a balance between domain regularization and human perceptual restoration. By taking the night-time image into our enhanced adversarial network, which employs structural similarity loss, we increase its visibility so that the driver can clearly see the road scene like they are driving during the day-time. Experimental results show the benefit of our approach for enhancing the visibility of the night-time view of the road, measured by the image quality metrics.

Keywords Deep learning · Traffic accidents · Generative adversarial networks (GAN)

1 Introduction

Traffic accident has many factors, one of which is the lack of road infrastructure such as inadequate road lighting. The World Health Organization (WHO) reported that road fatalities are one of the 10 leading causes of death in 2018. A study from our country's National Transportation Safety Committee revealed that accidents that occur at night and early morning have a portion of 34% of the total accidents that occur. Driving on roads that do not have inadequate street lighting infrastructure can increase the risk of traffic accidents. This risk can be reduced if road condition looks better and well perceived by the driver.

I. Ardiyanto (✉) · I. Soesanti · D. Cahya Qairawan
Department of Electrical Engineering and Information Technology,
Universitas Gadjah Mada, Jl. Grafika No. 2 Bulaksumur, Yogyakarta, Indonesia
e-mail: igi@ugm.ac.id

K. R. Ahmed et al. (eds.), *Deep Learning and Big Data for Intelligent Transportation*, Studies in Computational Intelligence 945,
https://doi.org/10.1007/978-3-030-65661-4_6

One possible way is to use computer vision to help the driver to see better on the road in dark conditions. The target task is to transform road images with the conditions of the night into the daytime so that the road and the objects around it look more clearly. In computer vision, such problem went into a translational task, and recent development suggests a technique called Generative Adversarial Networks (GAN) as the solution. GAN [1] is a framework that uses deep learning to create new data, including data in the form of images. GAN intents to learn two different networks; a generator and a discriminator networks. The generator part attemtps to attain synthetic data, while the discriminator part tries to distinguish the output of the generator with the true data. Creating a new data or image can also mean to convert the image from one domain to another while preserving the characteristic, which really fits to our problem above.

Several variants of GAN have been proposed by several researchers to accomplish specific tasks, and have shown convincing results in the field of image translation. For example, Iizuka et al. [2] employed GAN for image completion with globally and locally consistent adversarial training. There were also a work on Super Resolution using GAN (SRGAN) [3], transfering an image to the cartoon style using Domain Transfer Network [4], image inpainting using back propagation on a pretrained GAN [5], and image colorization from the grayscale input image [6]. Even, some works have been done for converting the day-time scene to the night-time one, such as [7, 8]. Nevertheless, changing images from night to day scenes remains a challenging problem.

To cope with the above problems, this research is expected to provide an option to reduce the risk of traffic accidents at night by translating road images with night to day conditions. In this research, we propose an enhanced adversarial network-based image translation. We modify its loss function by adding structural similarity loss into the existing. We expect the image quality improvement since the structural similarity considers the image perceptual terms, thus satisfying the human visual toward the translated night-time road scene. The translated images are then evaluated qualitatively and quantitatively to measure the visual prowess of the resulting output.

2 Image-to-Image Translation

One of plausible approach to convert the night-time scene into the day-time is by employing image-to-image translation. In the latest developments, Generative Adversarial Network (GAN) [1] becomes a standard approach for doing image translation tasks. Given X and Y representing two image domains, GAN is basically two networks (see Fig. 1), one is called Generator G and the other is Discriminator D. Both are competing against each other for learning generative models of random data distributions.

There are two possible ways for translating images using GAN, paired and unpaired. In paired style, the labeled/aligned image pairs in the same scenarios and conditions are often needed, and it is contrary for the unpaired one. In our cases, it

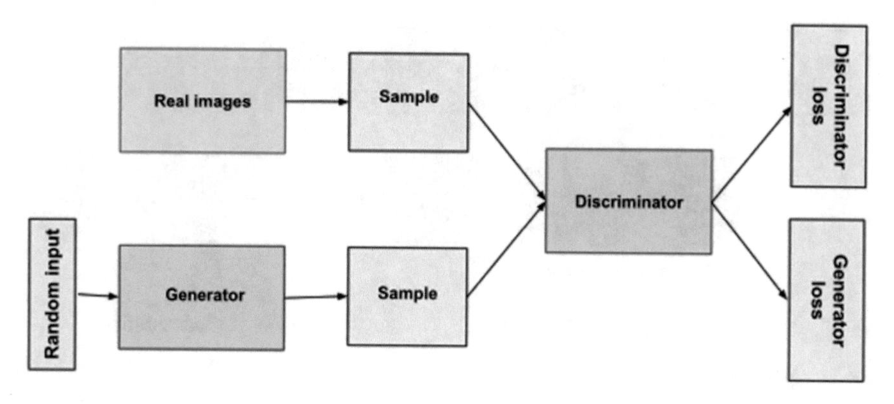

Fig. 1 GAN architecture

is impossible to get an exactly paired road images between the day-time and night-time, since the road scenes are dynamically changed second-by-second. Here we are focusing on the unpaired image translation methods, which was also employed by recent researches (e.g. [9–13]).

There are different methods for dealing with the unaligned image translations. CycleGAN [13] and DualGAN [14] are two of well-established algorithms for tackling it. CycleGAN employs two generators and two discriminators (G_X, D_X) and (G_Y, D_Y) so that $G_X : X \rightarrow Y$ and $G_Y : Y \rightarrow X$ with its corresponding discriminator D_X and D_Y (see Fig. 2). Both are competing for translating image from domain X to Y or the opposite. For the objective function, CycleGAN employs two losses, adversarial loss and cycle consistency loss, as follows

$$\mathcal{L} = \mathcal{L}_{adv} + \lambda \mathcal{L}_{cyc} \tag{1}$$

where

$$\begin{aligned} \mathcal{L}_{adv} = & \mathbb{E}_X[||(1 - D_Y(G_X(x)))^2||_1] \\ & + \mathbb{E}_Y[||(1 - D_X(G_Y(y)))^2||_1] \end{aligned} \tag{2}$$

$$\begin{aligned} \mathcal{L}_{cyc} = & \mathbb{E}_X[||G_Y(G_X(x)) - x||_1] \\ & + \mathbb{E}_Y[||G_X(G_Y(y)) - y||_1] \end{aligned} \tag{3}$$

where $x \in X$ and $y \in Y$, and λ is the tuning hyperparameter. From the above equations, the adversarial loss makes sure that the generated output will be on the appropriate domain, while the cycle consistency loss regularizes the input and output to be recognizably the same by mapping G_x and G_y such that it enforces $G_y(G_x(x)) \approx x$ and $G_x(G_y(y)) \approx y$.

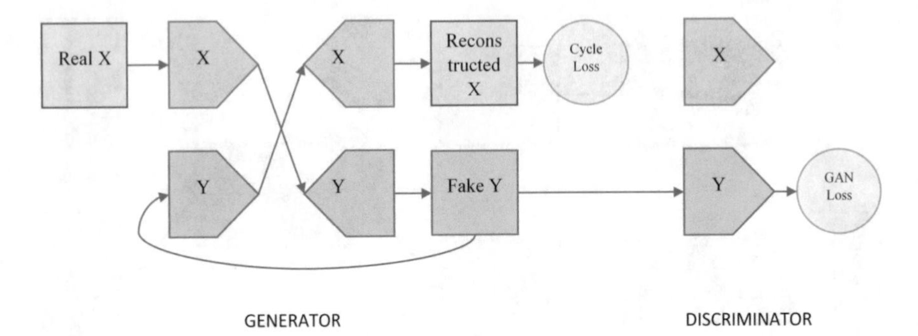

Fig. 2 CycleGAN architecture

On the other hand, DualGAN [14] utilizes reconstruction loss in place of cycle consistency loss, by measuring its distance with *mean averaged error*. Dual GAN measures likelihood of G_x belonging to target domain Y and *vice versa* using domain-adaptive GAN. Subsequently, it the loss function is described as follows

$$\begin{aligned}\mathcal{L}_{dual} =& \lambda_X||x - G_Y(G_X(x, z), z')|| \\ &+ \lambda_Y||y - G_X(G_Y(y, z'), z)|| \\ &- D_Y(G_Y(y, z')) - D_X(G_X(x, z))\end{aligned} \tag{4}$$

where z and z' are random noises, λ_X and λ_Y are constant parameters.

3 GAN with Structural Similarity Loss

One important thing to consider in translating night-time to day-time scene is the output quality. Empirically, we found that vanilla CycleGAN and DualGAN were failed to achieve a good output quality, especially in term of satisfying the human perception (i.e. visually pleasing). We hypothesize both algorithms are lacking of mechanism for measuring the restoration in their loss function. The restoration is very important since we deal with the task where the input has a low information (night-time scene) and is translated into the output with a high information (day-time scene). It is the opposite approach compared to the existing works which translate the day-time scene into the night one, e.g. [8]. In such problem setting, they have already had high information on the images, which are then translated into the lower one.

To cope with such problem, we adopt structural similarity [15] for perceiving the image quality. Structural similarity catches important perceptual terms, i.e. structural information, luminance, and contrast terms between two samples in each domain x and y, as follows,

$$SSIM(x, y) = l(x, y) * c(x, y) * s(x, y) \tag{5}$$

$$l(x, y) = \frac{2\mu_x\mu_y + k_1}{\mu_x^2 + \mu_y^2 + k_1} \tag{6}$$

$$c(x, y) = \frac{2\sigma_x\sigma_y + k_2}{\sigma_x^2 + \sigma_y^2 + k_2} \tag{7}$$

$$s(x, y) = \frac{\sigma_{xy} + k_3}{\sigma_x\sigma_y + k_3} \tag{8}$$

where μ and σ respectively denote mean intensity and standard deviation of the images, and $\{k_1, k_2, k_3\}$ are constants. Structural information $s(\cdot)$ employs a concept that the pixels which are adjacent spatially, have strong inter-dependencies. In the visual scene such as the road, it brings important information about the structure of the objects. Luminance term $l(\cdot)$ gives us an idea that image distortions tend to be less visible in bright regions, while contrast term $c(\cdot)$ explains that distortions become less visible in the area which has considerable activities, objects, or textures. In this case, an image with high structural similarity will be better suited to human perceptual judgements.

We basically inject the structural similarity as a weighted additional loss to Eqs. (1) and (4),

$$\mathcal{L}_{ssim} = \gamma(1 - SSIM(\hat{x}, \hat{y})) \tag{9}$$

where γ is a tuning hyperparameter, so that eq. (1) becomes

$$\mathcal{L} = \mathcal{L}_{adv} + \lambda\mathcal{L}_{cyc} + \mathcal{L}_{ssim}, \tag{10}$$

and eq. (4) evolves into

$$\begin{aligned} \mathcal{L}_{dual} = & \lambda_X||x - G_Y(G_X(x, z), z')|| \\ & + \lambda_Y||y - G_X(G_Y(y, z'), z)|| \\ & - D_Y(G_Y(y, z')) - D_X(G_X(x, z)) \\ & + \mathcal{L}_{ssim}. \end{aligned} \tag{11}$$

It makes balance between domain regularization and human perceptual restoration, so that the output images will have a nice visual pleasing while increases the amount of its information. Without such balance, the output image will have either a nice looking but miss some important information inside the image such as edge and gradient, or yield a good detail but have an ugly looking, as we will show in the results section.

4 Results

4.1 Dataset

To our knowledge, there is no public dataset dealing with the night-to-day road scene translations. It forced us to build our own dataset. Our dateset is made by recording video taken using a camera placed on a two-wheeled vehicle. The location of the video taking is in the area of Yogyakarta, Indonesia. The video is then processed to extract frames from the video. The frames are then selected based on predetermined criteria and grouped according to image conditions, day-time images and night-time images. Recording is done twice for the same route with different conditions (day-time and night-time). We use following criteria: image taken on asphalt road, no saturation, no over-exposure, and no too-dark-images. We collected a total 420 pair of day-night images which were resized to 256×256. Two images are considered as a pair if the day-time and the night-time images are taken in the same location, even with considerable translation and rotation. All images were then treated through a contrast enhancement before entering training and evaluation stage, as we will explain it later.

4.2 *Effects of SSIM Loss*

All results are acquired on Ubuntu PC with 32GB RAM and NVIDIA TitanXp graphic card. We first visually compare the performance of original CycleGAN and DualGAN with our proposed method. From Figs. 3 and 4, we are able to see the visual improvement of the output image when we add the structural similarity loss into the GANs. Objects, road marking, and even the background are clearer with firm edges and contrast. One interesting result on the third row, the rider shadow and vehicle light are even diminished in the translated image, which adds naturality of the output image. In our opinion, CycleGAN with additional SSIM-loss has the least image distortion and visually eye-pleasing.

We subsequently cross-check the above qualitative measurement with the measurable metrics. We employ three metrics, i.e. UQI, PSNR, and SSIM, for quantitatively measuring the performance of our proposed algorithm, as well as the quality of the output images, shown in Table 1. Universal Quality Index (UQI) [16] models the image distortion as a combination of linear correlation loss, luminance, and contrast distortion, which is almost similar to SSIM. The UQI metric Q is defined as follows

$$Q = \frac{\sigma_{xy}}{\sigma_x \sigma_y} \cdot \frac{2\bar{x}\bar{y}}{(\bar{x})^2 + (\bar{y})^2} \cdot \frac{2\sigma_x \sigma_y}{\sigma_x^2 + \sigma_y^2} \tag{12}$$

Fig. 3 Qualitative appearance of CycleGAN-based night-to-day

where x and y are the original daytime image and the output of our GAN system respectively, subject to

$$
\begin{aligned}
\bar{x} &= \frac{1}{N}\sum_{i=1}^{N} x_i, \qquad \bar{y} = \frac{1}{N}\sum_{i=1}^{N} y_i \\
\sigma_x^2 &= \frac{1}{N-1}\sum_{i=1}^{N}(x_i - \bar{x})^2, \qquad \sigma_y^2 = \frac{1}{N-1}\sum_{i=1}^{N}(y_i - \bar{y})^2 \\
\sigma_{xy} &= \frac{1}{N-1}\sum_{i=1}^{N}(x_i - \bar{x})(y_i - \bar{y}),
\end{aligned}
\tag{13}
$$

Fig. 4 Qualitative appearance of DualGAN-based night-to-day

with N be the total number of test images. All values are averaged from the 20% images of the dataset as the testing set.

As suggested by the original paper [16], it is advisable to compare statistical features locally in UQI. It is accomplished by employing sliding window of 8×8 to the image, and combining M set of those sliding windows for the overall quality, as follows

$$Q = \frac{1}{M} \sum_{i=1}^{M} Q_i. \tag{14}$$

For the PSNR, we leave the explanation to the reader since it is a widely known metric used in various image processing applications.

Table 1 Quantitative metric

Model	Avg. UQI	Avg. PSNR	Avg. SSIM
DualGAN	0.38	6.35	0.75
DualGAN + SSIM-Loss	0.42	6.87	0.76
CycleGAN	0.41	7.41	0.76
CycleGAN + SSIM-Loss	0.49	8.19	0.81

From Table 1, it can be clearly concluded that adding the structural similarity loss has a benefit for improving the image quality. The results show the distortions are reduced with the usage of structural similarity loss as well as the noise shown in the PSNR, thus improves the overall visibility quality of the images. It conforms with the visual results in Figs. 3 and 4.

4.3 Ablation Studies

Please be informed that the above results are taken using the most optimal image enhancement method in the preprocessing stage. For completeness, we now show the extended explanation to describe the effect of the image preprocessing to our proposed Night to Day algorithm. Particularly, we are interested on the searching of the most optimal contrast enhancement in our proposed solution. We hypothesized the contrast enhancement will increase the image entropy, which helped the convergence of the GAN system, thus, give a better day time output image perception.

Here we compare 10 contrast enhancement algorithms which were the state-of-the-art of low light image enhancement, i.e. Adaptive Integrated Nonlinear Enhancement (AINDANE) [17], Weighted Thresholded Histogram Equalization (WTHE) [18], Layered Difference Representation (LDR) [19], Adaptive Gamma Correction with Weighting Distribution (AGCWD) [20], Adaptive Gamma Correction for Image Enhancement (AGCIE) [21], Improved Adaptive Gamma Correction (IAGCWD) [22], Luminance Adaptation [23], Adaptive Image Enhancement (AIE) [24], Joint Histogram Equalization (JHE) [25], and Extended Exposure Fusion (SEF) [26]. Figures 5, 6, and 7 show the comparison on corresponding input images used in Figs. 3 and 4.

Tables 2 and 3 show the performance of each preprocessing technique towards our proposed algorithm. In general, the usage of the SSIM loss consistently yield a better image output quality. LDR [19] and IAGCWD [22] are the best choice for the pre-processing stage for our Night-to-Day image translation. Furthermore, a surprising result comes from the fact that even without preprocessing, our algorithm achieves a

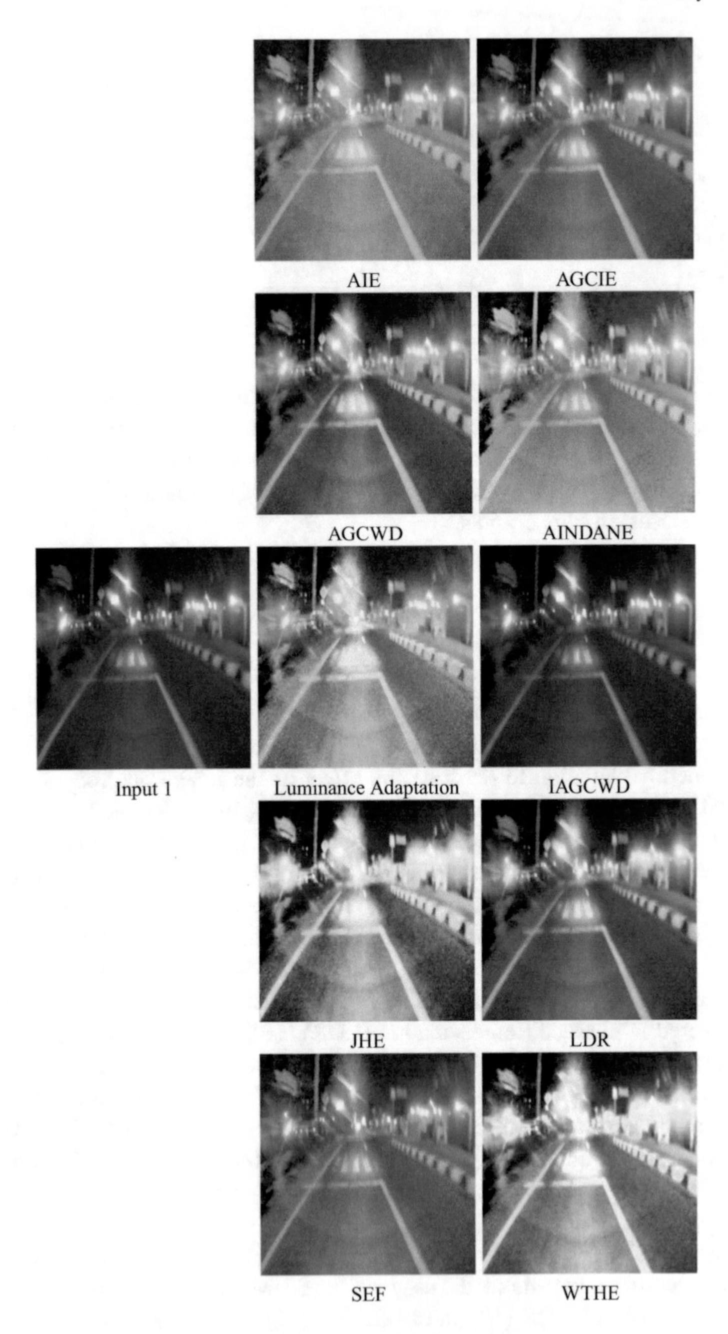

Fig. 5 Qualitative appearance of preprocessing effect on input 1

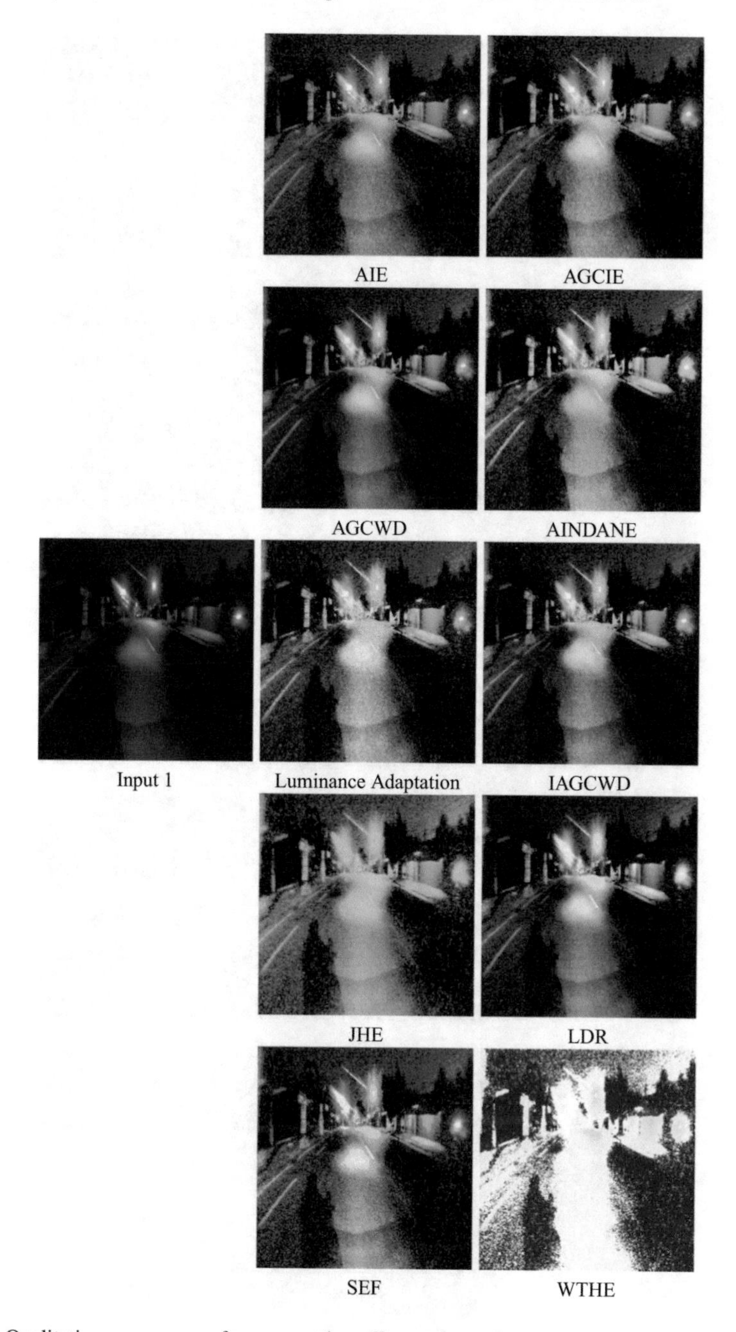

Fig. 6 Qualitative appearance of preprocessing effect on input 2

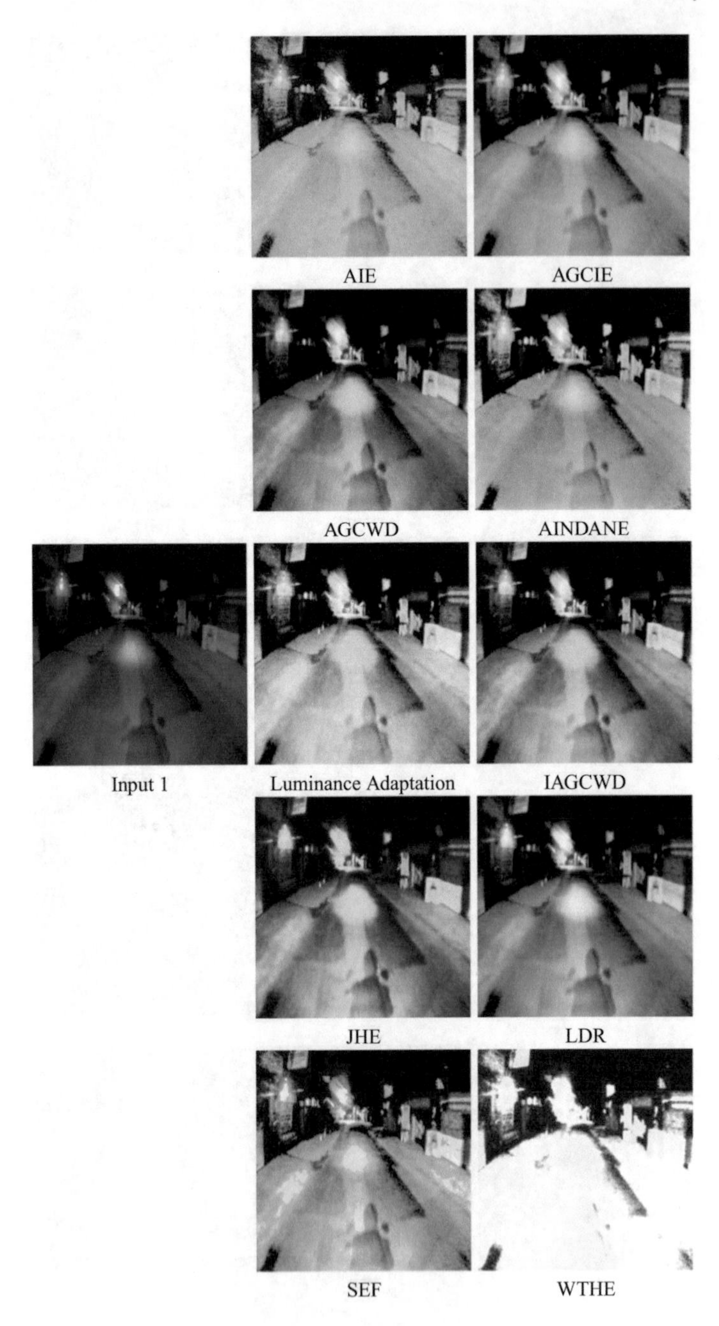

Fig. 7 Qualitative appearance of preprocessing effect on input 3

Table 2 Quantitative metrics of different preprocessing methods without SSIM loss

Preprocessing	DualGAN			CycleGAN		
	Avg. UQI	Avg. PSNR	Avg. SSIM	Avg. UQI	Avg. PSNR	Avg. SSIM
No Preprocessing	0.33	6.01	0.74	0.38	7.12	0.72
AIE	0.32	6.11	0.71	0.39	6.66	0.71
AGCIE	0.34	6.09	0.70	0.37	6.98	0.69
AGCWD	0.34	6.15	0.72	0.36	6.81	0.69
AINDANE	0.35	5.96	0.71	0.38	6.88	0.68
Luminance Adaptation	0.26	5.73	0.62	0.36	6.57	0.64
IAGCWD	**0.38**	6.31	**0.75**	**0.41**	7.40	**0.76**
JHE	0.36	6.29	0.73	0.37	7.36	0.71
LDR	0.36	**6.35**	0.74	0.39	**7.41**	0.75
SEF	0.31	6.17	0.70	0.33	7.03	0.69
WTHE	0.22	5.14	0.51	0.31	6.29	0.62

Table 3 Quantitative metrics of different preprocessing methods with additional SSIM loss

Preprocessing	DualGAN + SSIM			CycleGAN + SSIM		
	Avg. UQI	Avg. PSNR	Avg. SSIM	Avg. UQI	Avg. PSNR	Avg. SSIM
No Preprocessing	0.36	6.23	0.75	0.43	7.69	0.74
AIE	0.36	6.19	0.71	0.41	7.44	0.73
AGCIE	0.36	6.12	0.71	0.40	7.56	0.70
AGCWD	0.35	6.15	0.72	0.40	7.51	0.71
AINDANE	0.36	6.10	0.72	0.41	7.52	0.70
Luminance Adaptation	0.29	6.17	0.69	0.39	7.33	0.68
IAGCWD	**0.42**	6.61	**0.76**	**0.49**	8.08	**0.81**
JHE	0.38	6.42	0.74	0.44	7.78	0.74
LDR	0.41	**6.87**	0.75	0.48	**8.19**	0.80
SEF	0.36	6.26	0.70	0.41	7.67	0.76
WTHE	0.26	5.51	0.55	0.34	6.51	0.63

reasonable good result compared to the usage of several contrast enhancement methods. Especially for the PSNR, it seems the usage of contrast enhancement techniques unnecessarily bring additional noise to the input images, which explains the above phenomena.

5 Conclusion

In this research, we have shown the GAN variants can be used to translate the night-time road scenes to day-time conditions. We also introduce the additional structural similarity loss for improving the perceptual image quality. Both qualitative and quantitative measurements suggest ability of the proposed algorithm for transforming the night-time image into the day-time scene with considerable good quality. In the future, dataset enrichment is needed to improve the quality of output image further. The extendend choice of image preprocessing is also a good start for further improvement.

Acknowledgements The authors gratefully acknowledge NVIDIA for Titan XP donation used in this research.

References

1. Goodfellow I, Jean P-A, Mirza M, Xu B, David Warde-Farley, Sherjil Ozair, Aaron Courville, Yoshua Bengio (2014) Generative adversarial nets. Adv Neural Inf Process Syst 27:2672–2680
2. Iizuka S, Simo-Serra E, Ishikawa H (2017) Globally and locally consistent image completion. ACM Trans Graph (ToG) 36(4):1–4
3. Ledig C, Theis L, Huszár F, Caballero J, Cunningham A, Acosta A, Aitken A, Tejani A, Totz J, Wang Z, Shi W (2017) Photo-realistic single image super-resolution using a generative adversarial network. In: 2017 IEEE conference on computer vision and pattern recognition (CVPR), pp 105–114
4. Taigman Y, Polyak A, Wolf L (2017) Unsupervised cross-domain image generation. ArXiv, abs/1611.02200
5. Yeh RA, Chen C, Lim TY, Schwing AG, Hasegawa-Johnson M, Do MN (2017) Semantic image inpainting with deep generative models. In: 2017 IEEE conference on computer vision and pattern recognition (CVPR), pp 6882–6890
6. Cheng Z, Yang Q, Sheng B (2015) Deep colorization. In: Proceedings of the 2015 IEEE international conference on computer vision (ICCV), pp 415–423. ISBN 9781467383912
7. Liu M-Y, Breuel T, Kautz J (2017a) Unsupervised image-to-image translation networks. Adv Neural Inf Process Syst 30:700–708
8. Lin C, Huang S, Wu Y, Lai S (2020) Gan-based day-to-night image style transfer for nighttime vehicle detection. IEEE Trans Intell Transp Syst pp 1–13
9. Anoosheh A, Agustsson E, Timofte R, Van Gool L (2018) Combogan: unrestrained scalability for image domain translation. In: 2018 IEEE/CVF conference on computer vision and pattern recognition workshops (CVPRW), pp 896–8967
10. Choi Y, Choi M, Kim M, Ha J, Kim S, Choo J (2018) Stargan: unified generative adversarial networks for multi-domain image-to-image translation. In: 2018 IEEE/CVF conference on computer vision and pattern recognition, pp 8789–8797
11. Ignatov A, Kobyshev N, Timofte R, Vanhoey K, Van Gool L (2018) Wespe: weakly supervised photo enhancer for digital cameras. In: 2018 IEEE/CVF conference on computer vision and pattern recognition workshops (CVPRW), pp 804–809
12. Liu M-Y, Breuel T, Kautz J (2017b) Unsupervised image-to-image translation networks. Adv Neural Inf Process Syst 30:700–708
13. Zhu J, Park T, Isola P, Efros AA (2017) Unpaired image-to-image translation using cycle-consistent adversarial networks. In: 2017 IEEE international conference on computer vision (ICCV), pp 2242–2251

14. Yi Z, Zhang H, Tan P, Gong M (2017) Dualgan: unsupervised dual learning for image-to-image translation. In: 2017 IEEE international conference on computer vision (ICCV), pp 2868–2876
15. Wang Z, Bovik AC, Sheikh HR, Simoncelli EP (2004) Image quality assessment: from error visibility to structural similarity. IEEE Trans Image Process 13(4):600–612
16. Wang Z, Bovik AC (2002) A universal image quality index. IEEE Signal Process Lett 9(3):81–84
17. Tao L, Asari VK (2005) Adaptive and integrated neighborhood-dependent approach for nonlinear enhancement of color images. J Electron Imag 14(4):043006
18. Wang A, Ward RK (2007) Fast image/video contrast enhancement based on weighted thresholded histogram equalization. IEEE Trans Consum Electron 53(2):757–764
19. Lee C, Lee C, Kim C-S (2013) Contrast enhancement based on layered difference representation of 2d histograms. IEEE Trans Image Process 22(12):5372–5384
20. Huang S, Cheng F, Chiu Y (2013) Efficient contrast enhancement using adaptive gamma correction with weighting distribution. IEEE Trans Image Process 22(3):1032–1041
21. Rahman S, Rahman MM, Abdullah-Al-Wadud M, Al-Quaderi GD, Shoyaib M (2016) An adaptive gamma correction for image enhancement. Eurasip J Image Video Process 2016(1):35
22. Cao G, Huang L, Tian H, Huang X, Wang Y, Zhi R (2017) Contrast enhancement of brightness-distorted images by improved adaptive gamma correction. Comput Elect Eng 66:569–582
23. Fu Q, Jung C, Xu K (2018) Retinex-based perceptual contrast enhancement in images using luminance adaptation. IEEE Access 6:61277–61286
24. Wang W, Chen Z, Yuan X, Wu X (2019) Adaptive image enhancement method for correcting low-illumination images. Inf Sci 496:25–41
25. Agrawal S, Panda R, Mishro PK, Abraham A (2019) A novel joint histogram equalization based image contrast enhancement. J King Saud Univ Comput Inf Sci
26. Hessel C, Morel J-M (2020) An extended exposure fusion and its application to single image contrast enhancement. In: The IEEE winter conference on applications of computer vision, pp 137–146

Safer-Driving: Application of Deep Transfer Learning to Build Intelligent Transportation Systems

Ramazan Ünlü

Abstract Deep learning methods are widely used in the field of transportation as well as in many areas. These days, deep learning methods produce very successful results thanks to big data, where data collection and storage are much easier than in past years. However, it can be quite difficult to collect data or find the right data in some cases. On the other hand, even if the desired data is collected enough, training deep neural networks to learn from these data may not be comfortable in terms of hardware needed for ordinary users. In these cases, transfer learning can be an exit point, which is very popular in deep understanding, especially when the amount of data is limited. In this study, we have aimed to give comprehensive guidance on using a transfer learning strategy to create a detection model to establish a safer transportation system. More specifically, we have developed a classification model to alleviate accidents due to deer on the road. We have implemented VGG-16 convolutional neural network architecture as a part of the structured transfer learning framework. Computational results show that high accuracy performance can be achieved by implementing VGG-16 with the respectively small size of a dataset.

Keywords Intelligent transportation systems · Deep learning · Computer vision

1 Introduction

Deep learning allows computational models of multiple processing layers to learn and represent data with multiple levels of abstraction, imitating how the brain perceives and, thereby implicitly capturing complex structures of large-scale data. Deep learning is a sub-branch of machine learning and tries to learn hidden patterns from much larger data and complex structures. It is often used in many popular

R. Ünlü (✉)
Gumushane University, Baglarbası Mah, 29000 Gumushane, Merkez, Turkey
e-mail: ramazanunlu@gumushane.edu.tr

K. R. Ahmed et al. (eds.), *Deep Learning and Big Data for Intelligent Transportation*, Studies in Computational Intelligence 945,
https://doi.org/10.1007/978-3-030-65661-4_7

and classic domains of artificial intelligence, such as semantic parsing [9], transfer learning [44, 55], natural language processing [12, 29, 50], computer vision [18, 46], and many more.

Deep neural networks can be referred to as traditional neural networks with deeper structures. The history of deep neural networks goes back to the 1940s [35]. Nearly forty years later, with the discovery of the back-propagation algorithm by [39], it gained popularity again and many researchers started working in this field. However, due to the overfitting of the developed algorithms in training sets, limited computer power, etc., it lost its popularity again until the early 2000s. According to Guo et al. there are three main reasons behind the again booming of deep learning: the dramatically increased chip processing (e.g. using GPUs to process the data), the significantly lowered cost of computing hardware, and considerable advances in machine learning algorithms [11].

Deep learning consists of many different methods including, supervised and unsupervised, such as neural networks, hierarchical probabilistic models, and clustering. Deep learning has shown more successful results in many different areas, especially in the field of computer vision, than classical methods. Deep learning has provided great benefits in various domains due to cheaper data storage and faster data processing than in the past.

Many deep learning methods have been thoroughly reviewed and discussed in the literature [4–6, 18, 40]. Among those studies, Schmidhuber et al. give attention to essential milestones and technical contributions in a historical timeline format [40], while Bengio focused on challenges of deep learning studies and developed various feed-forward learning strategies [5]. Deep neural networks particularly have been shown to superior for computer vision tasks because they can automatically extract prominent features while jointly performing discrimination [25]. Various deep learning architectures have been effectively implemented by different researchers and achieved state-of-art accuracy scores in the recent ImageNet Large Scale Visual Recognition Challenge (ILSVRC) [10].

2 Deep Learning in Computer Vision

From the 1980s, convolutional neural networks (CNN) have been frequently used in computer vision tasks and have achieved prominent results in different challenges. But except for some applications, the increase in computer power, the development of more advanced algorithms, and the facilitation of access to enormous amounts of labeled data from the beginning of the 2010s have experienced the renaissance in the field of computer vision.

Image classification is one of the popular applications in computer vision, and CNN has shown tremendous performance on those tasks. Image classification can be defined as the task of categorizing images into correct classes and is one of the main tasks in computer vision. Also, it can be thought of as the basis of other computer vision tasks such as localization [2], detection [54], and segmentation [3].

The crucial part of image processing is the extraction of features from the images. In recent years, deep learning models have used different strategies through multiple layers of nonlinear information processing to extract prominent features as well as for pattern analysis and classification [27, 38]. By implementing such strategies, CNN has become the leading architecture for most of the image recognition, classification, and object detection tasks [26]. Despite some of the early successful application of CNN [28, 41], they have produced the state-of-art results by fueling GPUs, larger datasets, and better algorithms [13, 24, 42, 52]. A few advances such as the first GPU implementation [7] and the first application of max pooling strategy have contributed to CNN's popularity [37].

The milestone of deep learning algorithms in image classification problems is the success in the ImageNet Large Scale Visual Recognition Challenge (ILSVRC). Krizhevsky et al. used a deep convolutional neural network to classify approximately 1.2 million images into 1000 classes [24]. After this achievement, deep convolutional neural networks have been frequently used in a different version of ILSVRC with the aim of image classification tasks [42, 43, 52]. Some of the researchers have also developed some state-of-art CNN architectures [17, 32, 43, 51], non-linear activation functions [19, 48], regularization techniques [20, 51], and optimization methods [16, 24]. Researchers still work to enhance a better framework to achieve better performance in various image classification problems.

Deep learning methods, including deep convolutional neural networks, are increasing their popularity day by day and are still frequently used in many areas, including transportation systems. It will undoubtedly continue to be used today and in the future in the development of safer transportation systems.

2.1 CNNs Architecture

CNNs are feedforward networks in that information goes from in one direction and similar to standard neural network the weight parameters are optimized by using the back-propagation algorithm. There are many different architectures of CNNs; however, in general, they consist of two main layers that differ from traditional neural networks which are called convolutional and pooling layers. These two main components of CNNs are followed by fully connected layers which are standard feedforward neural networks. Figure 1 illustrates a typical CNN architecture.

Without needing a manual feature extraction process, an image is directly going to convolutional layers as a raw input. Thereafter, one or several fully connected layers are fed by automatically extracted features. Finally, the last layer, or called the output layer, is used to assign an image to the predefined classes. Besides this standard architecture, various studies have proposed different architecture to improve the accuracy score of the image classification task.

Convolutional Layer

The convolutional layer is a module to extract the features, and thus the process learns feature representations of the input image. The convolutional process works

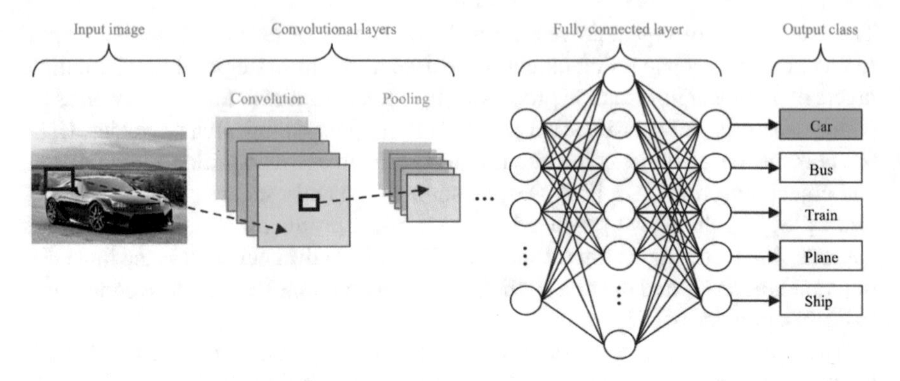

Fig. 1 CNN image classification pipeline [38]

based on moving a $f \times f$ predefined filter across the image [26]. Inputs, or pixels of the image, are convolved with weights aiming to compute a new feature map, and the feature map is sent through a nonlinear activation function. Nonlinear functions are implemented for the revealing of nonlinear features. The sigmoid and hyperbolic tangent functions were commonly used in many different applications; bu recently, rectified linear units (ReLUs) have become popular in the image classification process [26]. By inventing the effect of activation functions, many researchers have focused on the development of state-of-art activations functions [21, 23, 31, 36, 53]. The following Fig. 2 illustrates a simple convolutional operation with the stride of 2 meaning that the filter will be move from left to right and from the top the down with step size 2.

Pooling Layer

The pooling layer is used to aggregate information within each small region of the input features channels and reduce the size of the feature map. Similar to convolutional operation, a filter is used across the image, but without any weights. In literature, there are two main pooling operations which are Max pooling and Average pooling. Simply, the max-pooling operator takes the maximum value of the region

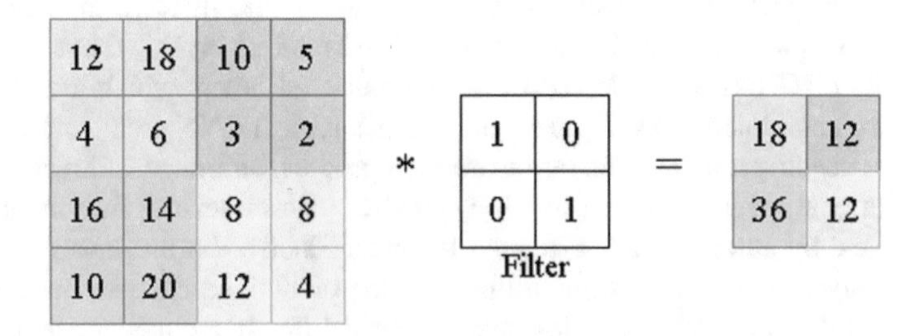

Fig. 2 Convolutional operation with the stride of 2

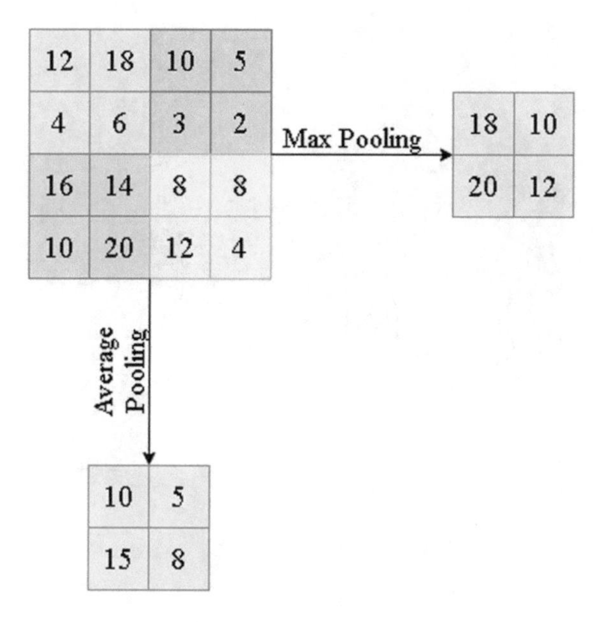

Fig. 3 Average and maximum pooling

that corresponds to filter size, and the average pooling operator takes the average value of the region that corresponds to filter size. Figure 3 illustrates the average and maximum pooling operators.

In general, convolutional operation and pooling operation is combined to create a new feature map then a nonlinear function is applied to reveal nonlinear hidden features. Thereafter, that new feature map is used as the input of a fully connected layer.

3 Transfer Learning

Training a deep neural network from scratch is not an easy task because of various reasons. First, deep architecture usually needs a dataset of sufficient size. And, it sometimes is not an easy task to collect or find enough data. In the case of availability of the required data is found, reaching the convergence can take a long time and needs powerful computational capability. Even if a large enough dataset available and convergence does not take a long time, it is proved that using pre-trained weights can provide great benefits [1, 34]. Yosinski et al. [49] have proved that using a pre-trained network, even for a different task, can perform better performance than training from scratch. In the literature, we can find various pre-trained deep convolutional neural networks showed superior performance in different tasks. To apply one of these models, one needs to decide what label retrained in order to achieve the highest

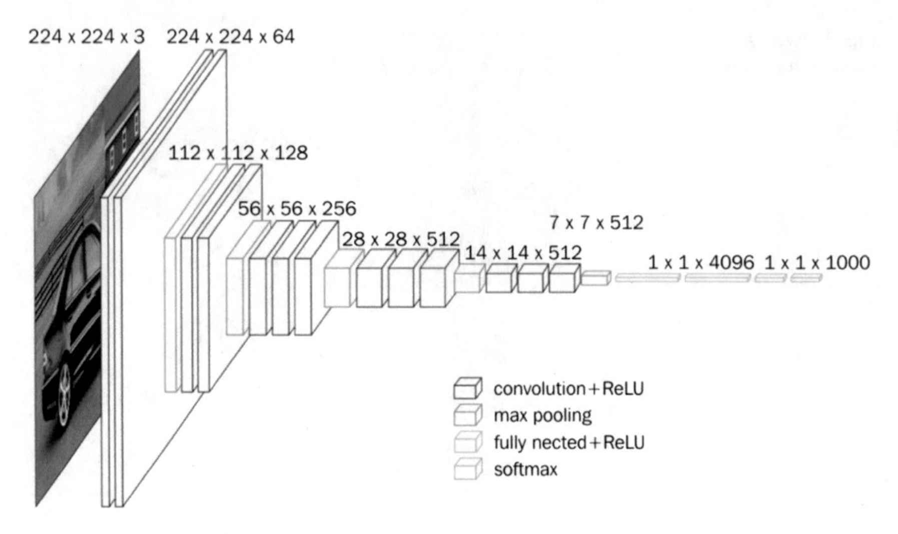

Fig. 4 VGG-16 architecture

performance. Usually, fine-tuning the higher level of a pre-trained network is better than fine-tuning the lower level part, since lower levels of pretrained neural network consist of more generic features.

In different deep learning libraries, we can use many different pre-trained neural networks. Among them, we have chosen VGG-16 architecture and reused the fine-tuned weights in the ImageNet dataset. VGG-16 architecture is proposed by Visual Geometry Group (VGG) from the University of Oxford [42]. VGG-16 model is composed of 16 weights layers and became popular due to its achievement of 92.7% top-5 test accuracy. The following Fig. 4 shows the architecture of the VGG-16 model. As can be seen, the model is structured such that input images will be the size of 224 $\times$ 224. So, we have resized each image based on these measures.

4 Deep Learning in Transportation

Intelligent transportation systems (ITSs) will be the main component of the smart cities of the future. However, that may require a tremendous and efficient data analytics process. Due to the nature of the transportation systems, the data that can be obtained from this area is very large and tends to be complex. The processing of big data requires both high computing power and effective algorithms. From the context of Intelligent Transportation Systems (ITSs), data can be collected from many sources such as mobile phones, vehicles, and CCTV cameras, and it is inevitable to obtain data in many different structures such as audio, images, and numeric.

ITSs, which will be indisputably one of the most important components of future smart cities, are now actively used in some areas, such as autonomous cars. And this use will increase more and more in line with advances in transportation infrastructures and technologies, as urbanization has shown extraordinary growth today.

For example, traffic congestion is a very important problem in terms of time loss, productivity, air pollution, and wasted energy in residential centers where urbanization is high. Today such problems can be eliminated thanks to high-end data processing technologies, at least brought to a minimum level. For the purpose of aggregating data, a tremendous number of sensors have been deployed and continued to be used to provide streaming real-time data. Thanks to this, advanced technologies that can process this data have been developed and are still being enhanced. Thus, intelligent transportation systems are developing a little more every single day. With the advent of computational power and developed state-of-art methods, data processing technologies can handle real-time, complex, and heterogeneous data for a better real-time decision support system.

Deep Learning has been effectively used in many different studies in the field of transportation problems, such as traffic flow prediction [30], traffic signal control [33], vehicle or object detection [22], travel demand prediction [15], traffic incident processing [8], autonomous driving [22], driver behaviors [14] and many more [45]. In these studies, many artificial intelligence methods such as multilayer perceptron, convolutional neural networks, recurrent neural networks, etc. have been used for years. Wang et al. [47] gives a comprehensive survey and summarized deep learning methodologies applied to studies related to the transportation field as shown in Fig. 5.

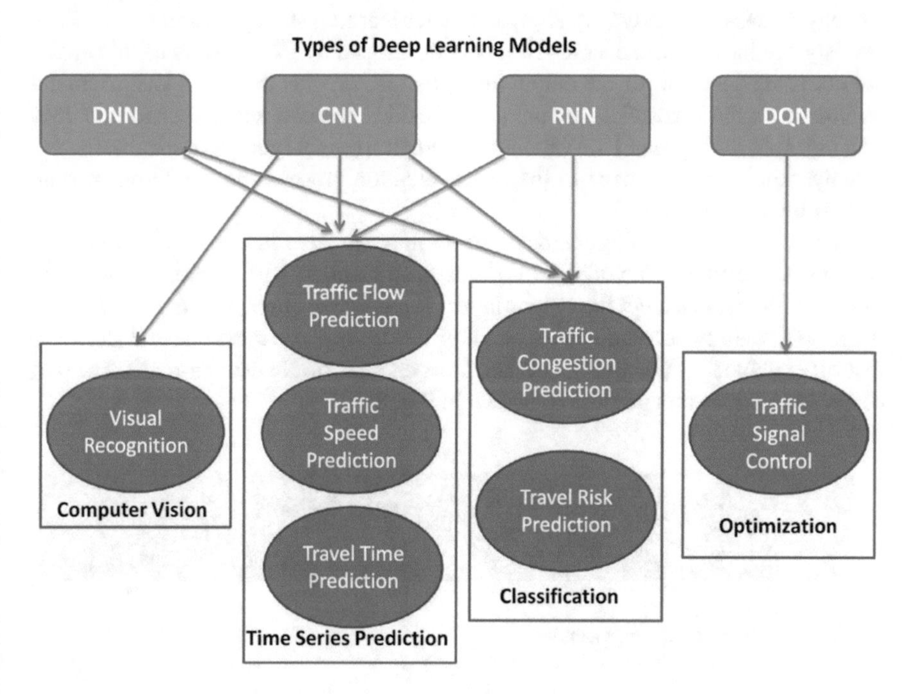

Fig. 5 Deep learning methodologies used in transportation studies [47]

5 Case Study: Using Transfer Learning to Create Object Classification Model

In this section, we have created a simple but comprehensive object classification model to ensure a safer transportation system. As it is known, there are many accidents caused by animals on some road routes. For example, in the USA, deer, who suddenly set off, cause many accidents. These accidents endanger the safety of transportation systems due to loss of life and loss of time. To avoid this, let's assume that we want to develop a deep learning-based system, and thanks to that system, if there's a deer on the road, we want to identify it as soon as possible.

The first stage for the development of such a model is to collect data and it is possible to perform data collection in many different ways. For example, video can be shot by inserting a camera inside the vehicle, and then parsed video frames to suit the classification problem. In this study, we have collected the necessary input data, namely images, from open sources such as images.google.com. The total number of images collected is 526 consisting of 250 are images of a deer on the road, the remaining 250 images without any deer on the road. Figure 6 shows some of these images.

To develop an effective classification model, we can design a model from scratch or can adapt a successful model used in literature for our problem. We explained the details of this strategy in the Transfer Learning section in detail. Our preferred strategy in this case study is to apply a transfer learning strategy. To implement this strategy, we have used the VGG-16 model developed by [42]. Layers of this model are not trained except for the output layer, that is, weights that the VGG-16 model optimized on the ImageNet data are used. Besides, the data set is divided into 75% training, 15% testing, and 15% validation to create a reliable and generalizable model. Mainly, this is the first stage of the model creation process and the corresponding code is illustrated in Fig. 7.

The next step is training the data. To do this, we have first load predetermined weights of the VGG-16 model from Keras deep learning API. Thereafter, we need to freeze all layers except the output layer. The number of neurons in the last layer is fixed for the ImageNet database and it is 1000. Thus, we must revise it as two to make it suitable for our experiment (i.e., Deer on road vs. No deer on-road). The part of the code serves this purpose as shown in Fig. 8.

Fig. 6 Illustration of some data samples

```
import random
import os
import numpy as np
import keras
import matplotlib.pyplot as plt
from matplotlib.pyplot import imshow
from keras.preprocessing import image
from keras.applications.imagenet_utils import preprocess_input
from keras.models import Sequential
from keras.layers import Dense, Dropout, Flatten, Activation
from keras.layers import Conv2D, MaxPooling2D
from keras.models import Model

root = 'data/SpringerData'
train_split, val_split = 0.60, 0.20
categories = [x[0] for x in os.walk(root) if x[0]][1:]

#create function to Load the data
def get_image(path):
    img = image.load_img(path, target_size=(224, 224))
    x = image.img_to_array(img)
    x = np.expand_dims(x, axis=0)
    x = preprocess_input(x)
    return img, x

data = []
for c, category in enumerate(categories):
    images = [os.path.join(dp, f) for dp, dn, filenames
              in os.walk(category) for f in filenames
              if os.path.splitext(f)[1].lower() in ['.jpg','.png','.jpeg']]
    for img_path in images:
        img, x = get_image(img_path)
        data.append({'x':np.array(x[0]), 'y':c})

# count the number of classes
num_classes = len(categories)


random.shuffle(data)

idx_val = int(train_split * len(data))
idx_test = int((train_split + val_split) * len(data))
train = data[:idx_val]
val = data[idx_val:idx_test]
test = data[idx_test:]


x_train=[],y_train=[],x_val=[],y_val=[],x_test=[],y_test=[]

for i in range (len(train)):
    x_train.append(train[i]['x'])

for i in range (len(train)):
    y_train.append(train[i]['y'])

for i in range (len(val)):
    x_val.append(val[i]['x'])

for i in range (len(test)):
    x_test.append(test[i]['x'])

for i in range (len(test)):
    y_test.append(test[i]['y'])

# Normalizing data
x_train = np.array(x_train).astype('float32') / 255.
x_val = np.array(x_val).astype('float32') / 255.
x_test = np.array(x_test).astype('float32') / 255.

# Convert Labels to one-hot vectors
y_train = keras.utils.to_categorical(y_train, num_classes)
y_val = keras.utils.to_categorical(y_val, num_classes)
y_test = keras.utils.to_categorical(y_test, num_classes)
```

Fig. 7 Python codes to prepare datasets for the train, test, and validation process

The last step is training the model. To do this, we can use "*model.fit*" function. The model will learn the hidden pattern from training data and parameters are optimized in the validation set. The following part of the code is shown in Fig. 9 works for this purpose.

Now, let's look at the results of the model. We can first look at the loss values of the model in training and test set over the number of epochs. The following Fig. 10 illustrates the loss values of the VGG-16 Model. One needs to note that, VGG-16 model can be highly complex for this case study because of the size of the dataset,

```
VGG19=keras.applications.mobilenet_v2.MobileNetV2(weights='imagenet', include_top=True)
inp = VGG19.input
new_classification_layer = Dense(num_classes, activation='softmax')
out = new_classification_layer(VGG19.layers[-2].output)
model_VGG19 = Model(inp, out)

for l, layer in enumerate(model_VGG19.layers[:-1]):
    layer.trainable = False

for l, layer in enumerate(model_VGG19.layers[-1:]):
    layer.trainable = True

model_VGG19.compile(loss='categorical_crossentropy',
            optimizer='adam',
            metrics=['accuracy'])

model_VGG19.summary()
```

Fig. 8 Python codes to revise the VGG-16 model for the proposed task

```
batch_size=4
epochs=100

history_VGG19 = model_VGG19.fit(x_train, y_train,
                        epochs=epochs,
                        batch_size=batch_size,
                        validation_data=(x_val, y_val))
```

Fig. 9 Python codes to train and validate datasets

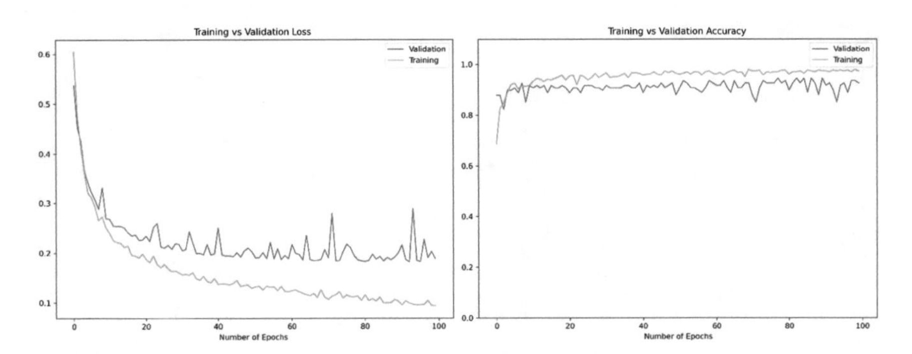

Fig. 10 Loss and accuracy performance of the model in the train and validation set

whereas the reason behind preferring this model is to give comprehensive guidance on using transfer learning methodology.

In a classification model, the expected change is that loss values decrease in both training and validation set in the number of epochs in parallel. This value may increase in the validation set after a certain point and this means that the model is overfitting in the validation set. To prevent this situation, dropout or various regularization methods can be used first. Another option is to create more validation samples. To do this, it is needed to conduct a more thorough and rigorous data collection process.

On the other hand, another strategy, called early stopping, is commonly used through the model creation process. The fundamental principle behind this strategy is stopping the model at the point where the model performs best and the loss values cannot be further reduced. Thereafter, the optimum weights are reloaded and used for the final prediction model. The created model can be tested in a train and validation

```
loss, accuracy = model_VGG16.evaluate(x_test, y_test, verbose=0)

print('Test loss:', loss)
print('Test accuracy:', accuracy)

loss, accuracy = model_VGG16.evaluate(x_val, y_val, verbose=0)

print('Val loss:', loss)
print('Val accuracy:', accuracy)

loss, accuracy = model_VGG16.evaluate(x_train, y_train, verbose=0)

print('Train loss:', loss)
print('Train accuracy:', accuracy)
```

Fig. 11 Python codes to calculate the loss and accuracy of the model in the train, test, and validation set

set. Also, to see how the created model performs on an unseen dataset, the model can be used in test data. The part of the code to calculate the cost and accuracy of the model in training, validation, and test set is shown in Fig. 11.

The performance of the VGG-16 model for this case study is shown in Table 1. Accuracy and loss values alone can be misleading in some cases. To avoid this misorientation, it is necessary to look at different evaluation metrics by looking at the confusion matrix. The confusion matrix is a $k \times k$ matrix that shows how many of the data samples are assigned to the correct class. Figure 12 illustrates the confusion matrix of the VGG-16 model for this case study.

In this figure, we have shown the normalized confusion matrix along with the number of assigned data classes. Based on these numbers, a total of 108 images are used to test the trained model. 53 of these images have a deer and 55 images do not have a deer. Also, 6 of 53 "Deer on Road" images are assigned to the wrong classes. In other words, these 6 "Deer on Road" images are predicted as "Deer not on Road" images. Similarly, there are 55 "Deer not on Road" images and three of them are predicted as "Deer on Road" pictures.

From these numbers, we can calculate several evaluation metrics that can give better guidance. These are called Sensitivity, Specificity, and G-Mean. The following Table 2 gives the formulation of these evaluation metrics and the corresponding performance of the implemented VGG-16 model.

Sensitivity is defined as the proportion of true positives samples correctly assigned, such as the percentage of "Deer on Road" images are classified as having a deer on road. The sensitivity score of the trained model is 0.940, so it can be concluded the

Table 1 Accuracy and loss performance of the model in train, test, and validation set

Dataset	Performance
Test loss	0.194
Test accuracy	0.916
Validation loss	0.187
Validation accuracy	0.925
Training loss	0.086
Training accuracy	0.981

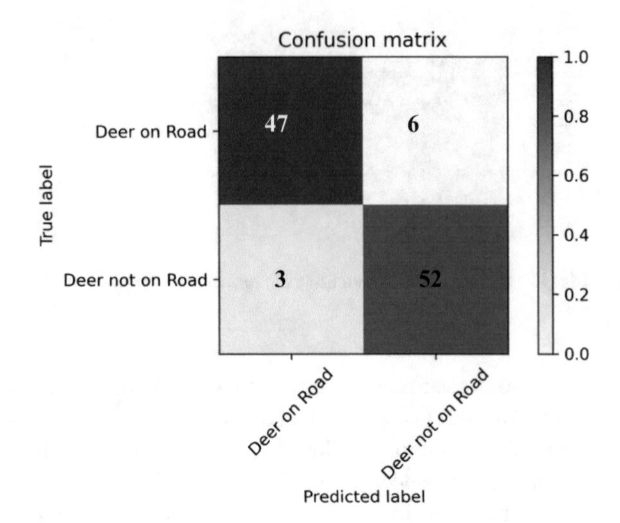

Fig. 12 Confusion matrix of the model in the test set

Table 2 The formulation of evaluation metrics and corresponding results

Evaluation metrics	Formulation	Results
Sensitivity	$\frac{TP}{TP+FN}$	0.940
Specificity	$\frac{TN}{TN+FP}$	0.896
G-Mean	$\sqrt{Sensitivity \times Specificity}$	0.917

trained model shows a good performance on classifying the true positives samples. In turn, the specificity score can be defined as the proportion of negative samples correctly assigned, such as the percentage of "Deer not on Road" images are classified as no deer on road. The specificity score of the trained model is 0.896, it is not as good as its sensitivity score, but well enough. Finally, the G-Mean score is the geometric average of sensitivity and specificity and gives a better intuition about how the model performs both in sensitivity and specificity. Our VGG-16 model performs well in terms of G-Mean with a score of 0.917. Thus, the developed model performs well enough to detect both positive and negative samples.

6 Conclusion

Deep learning systems are often used in transportation as well as in many areas. Especially following the concept of smart cities, artificial intelligence algorithms, especially deep learning systems, have been used to solve many problems and come up with more effective and faster solutions to these problems.

In this study, we have utilized a deep transfer learning framework to show how we can create a deep learning-based solution to provide a safer transportation system. To create such a solution for a specific problem, we can create a model from scratch or apply a pre-trained model. To create a model from scratch, we first need enough number of data samples and computational power. Usually, deep learning models perform better than traditional machine learning models if we can feed them with big data. And, to process such a big data, we need to enough computational power (i.e. GPU). The availability of enough number of data and computational power might not be easy. If this is the case, we can use a pre-trained model to get the best possible performance with a small amount of data. One can find various pre-trained deep learning models and they have achieved tremendous performance in different big datasets. VGG-16 convolutional neural network architecture is one of those models and achieved a prominent performance on the ImageNet database. We have applied this model into our case study which is about detecting a deer on road. VGG-16 model is available in Keras deep learning API and one can find other well-known pre-trained models shown in Table 3 in which the top-1 and top-5 accuracy refers to the model's performance on the ImageNet validation dataset and depth refers to the topological depth of the network. This includes activation layers, batch normalization layers, etc.

Through the model creation process, we first need to collect data samples. In our case, the data samples are the road images either having or not having a deer. We have collected these images from images.google.com. Thereafter, we need to preprocess these images. VGG-16 model is designed for the input image with a size of 224 $\times$ 224. So, we resized each image as being 224 $\times$ 224 pixels. Then, the preprocessed data samples are split as train, test, and validation with the rate of %60, %20, and %20, respectively.

Finally, we have trained the VGG-16 model with the train set and weight parameters are optimized in the validation set. By doing this, the weights are optimized based on accuracy metrics. The created model is used in the test set to see how the trained model performs in an unseen dataset. So, Our trained model gives 0.916, 0.925, and 0.981accuracy rates for the test, validation, and training set, respectively.

In this case study, we aimed to provide a simple but comprehensive guide on how to use deep transfer learning models in the field of transportation systems. The created model can be enhanced by collecting more data samples and by creating a model from scratch for the proposed task specifically. By doing this, we can come up with a model that can be used in real-world applications in a better sense.

Table 3 Pretrained models in Keras deep learning library

Model	Size (MB)	Top-1 accuracy	Top-5 accuracy	Parameters	Depth
Xception	88	0.790	0.945	22,910,480	126
VGG-16	528	0.713	0.901	138,357,544	23
VGG19	549	0.713	0.900	143,667,240	26
ResNet50	98	0.749	0.921	25,636,712	–
ResNet101	171	0.764	0.928	44,707,176	–
ResNet152	232	0.766	0.931	60,419,944	–
ResNet50V2	98	0.760	0.930	25,613,800	–
ResNet101V2	171	0.772	0.938	44,675,560	–
ResNet152V2	232	0.780	0.942	60,380,648	–
InceptionV3	92	0.779	0.937	23,851,784	159
InceptionResNetV2	215	0.803	0.953	55,873,736	572
MobileNet	16	0.704	0.895	4,253,864	88
MobileNetV2	14	0.713	0.901	3,538,984	88
DenseNet121	33	0.750	0.923	8,062,504	121
DenseNet169	57	0.762	0.932	14,307,880	169
DenseNet201	80	0.773	0.936	20,242,984	201
NASNetMobile	23	0.744	0.919	5,326,716	–
NASNetLarge	343	0.825	0.960	88,949,818	–
EfficientNetB0	29	–	–	5,330,571	–
EfficientNetB1	31	–	–	7,856,239	–
EfficientNetB2	36	–	–	9,177,569	–
EfficientNetB3	48	–	–	12,320,535	–
EfficientNetB4	75	–	–	19,466,823	–
EfficientNetB5	118	–	–	30,562,527	–
EfficientNetB6	166	–	–	4,326,514	–
EfficientNetB7	256	–	–	66,658,687	–

References

1. Ahmed A, Yu K, Xu W, Gong Y, Xing E (2008) Training hierarchical feed-forward visual recognition models using transfer learning from pseudo-tasks. In: European conference on computer vision. Springer, pp 69–82
2. Asghar K, Habib Z, Hussain M (2017) Copy-move and splicing image forgery detection and localization techniques: a review. Aust J Forensic Sci 49:281–307
3. Badrinarayanan V, Kendall A, Cipolla R (2017) Segnet: a deep convolutional encoder-decoder architecture for image segmentation. IEEE Trans Pattern Anal Mach Intell 39:2481–2495
4. Bengio Y (2009) Learning deep architectures for AI. Found trends® Mach Learn 2:1–127
5. Bengio Y (2013) Deep learning of representations: looking forward. In: International conference on statistical language and speech processing. Springer, pp 1–37

6. Bengio Y, Courville A, Vincent P (2013) Representation learning: A review and new perspectives. IEEE Trans Pattern Anal Mach Intell 35:1798–1828
7. Chellapilla K, Puri S, Simard P (2006) High performance convolutional neural networks for document processing
8. Chen Q, Song X, Yamada H, Shibasaki R (2016) Learning deep representation from big and heterogeneous data for traffic accident inference. In: Thirtieth AAAI conference on artificial intelligence
9. Dauphin YN, Hakkani-Tur DZ, Tur G, Heck LP (2015) Deep learning for semantic parsing including semantic utterance classification
10. Deng J, Dong W, Socher R, Li L-J, Li K, Fei-Fei L (2009) Imagenet: A large-scale hierarchical image database. In: 2009 IEEE conference on computer vision and pattern recognition. IEEE, pp 248–255
11. Deng L (2014) A tutorial survey of architectures, algorithms, and applications for deep learning. APSIPA Trans Signal Inf Process 3
12. Deng L, Liu Y (2018) Deep learning in natural language processing. Springer
13. Deng L, Yu D (2014) Deep learning: methods and applications. In: Foundations and trends in signal processing series book, vol 20
14. Dong W, Li J, Yao R, Li C, Yuan T, Wang L (2016) Characterizing driving styles with deep learning. arXiv Prepr arXiv160703611
15. Geng X, Li Y, Wang L, Zhang L, Yang Q, Ye J, Liu Y (2019) Spatiotemporal multi-graph convolution network for ride-hailing demand forecasting. In: Proceedings of the AAAI conference on artificial intelligence, pp 3656–3663
16. Glorot X, Bengio Y (2010) Understanding the difficulty of training deep feedforward neural networks. In: Proceedings of the thirteenth international conference on artificial intelligence and statistics. pp 249–256
17. Gong Y, Wang L, Guo R, Lazebnik S (2014) Multi-scale orderless pooling of deep convolutional activation features. In: European conference on computer vision. Springer, pp 392–407
18. Guo Y, Liu Y, Oerlemans A, Lao S, Wu S, Lew MS (2016) Deep learning for visual understanding: a review. Neurocomputing 187:27–48
19. He K, Zhang X, Ren S, Sun J (2016) Deep residual learning for image recognition. In: Proceedings of the IEEE conference on computer vision and pattern recognition. pp 770–778
20. Hinton GE, Srivastava N, Krizhevsky A, Sutskever I, Salakhutdinov RR (2012) Improving neural networks by preventing co-adaptation of feature detectors. arXiv Prepr arXiv12070580
21. Hou L, Samaras D, Kurc TM, Gao Y, Saltz JH (2017) ConvNets with smooth adaptive activation functions for regression. Proc Mach Learn Res 54:430
22. Huval B, Wang T, Tandon S, Kiske J, Song W, Pazhayampallil J, Andriluka M, Rajpurkar P, Migimatsu T, Cheng-Yue R (2015) An empirical evaluation of deep learning on highway driving. arXiv Prepr arXiv150401716
23. Kakuda K, Enomoto T, Miura S (2018) Nonlinear activation functions in CNN based on fluid dynamics and its applications. Comput Model Eng Sci 118:1–14
24. Krizhevsky A, Sutskever I, Hinton GE (2012) Imagenet classification with deep convolutional neural networks. In: Advances in neural information processing systems, pp 1097–1105
25. LeCun Y (2012) Learning invariant feature hierarchies. In: European conference on computer vision. Springer, pp 496–505
26. LeCun Y, Bengio Y, Hinton G (2015) Deep learning. Nature 521:436
27. LeCun Y, Boser B, Denker JS, Henderson D, Howard RE, Hubbard W, Jackel LD (1989) Backpropagation applied to handwritten zip code recognition. Neural Comput 1:541–551
28. LeCun Y, Bottou L, Bengio Y, Haffner P (1998) Gradient-based learning applied to document recognition. Proc IEEE 86:2278–2324
29. Li H (2017) Deep learning for natural language processing: advantages and challenges. Natl Sci Rev
30. Li J, Peng H, Liu L, Xiong G, Du B, Ma H, Wang L, Bhuiyan MZA (2018) Graph CNNs for urban traffic passenger flows prediction. In: 2018 IEEE SmartWorld, ubiquitous intelligence and computing, advanced and trusted computing, scalable computing and communications, cloud and big data computing, internet of people and smart City innovation (SmartWorld/SCALCOM/UIC/ATC/CBDCom/IOP/SCI). IEEE, pp 29–36

31. Liew SS, Khalil-Hani M, Bakhteri R (2016) Bounded activation functions for enhanced training stability of deep neural networks on visual pattern recognition problems. Neurocomputing 216:718–734
32. Lin M, Chen Q, Yan S (2013) Network in network. arXiv Prepr arXiv13124400
33. Nishi T, Otaki K, Hayakawa K, Yoshimura T (2018) Traffic signal control based on reinforcement learning with graph convolutional neural nets. In: 2018 21st International conference on intelligent transportation systems (ITSC). IEEE, pp 877–883
34. Oquab M, Bottou L, Laptev I, Sivic J (2014) Learning and transferring mid-level image representations using convolutional neural networks. In: Proceedings of the IEEE conference on computer vision and pattern recognition, pp 1717–1724
35. Pitts W, McCulloch WS (1947) How we know universals the perception of auditory and visual forms. Bull Math Biophys 9:127–147
36. Qian S, Liu H, Liu C, Wu S, San Wong H (2018) Adaptive activation functions in convolutional neural networks. Neurocomputing 272:204–212
37. Ranzato M, Huang FJ, Boureau Y-L, LeCun Y (2007) Unsupervised learning of invariant feature hierarchies with applications to object recognition. In: 2007 IEEE conference on computer vision and pattern recognition. IEEE, pp 1–8
38. Rawat W, Wang Z (2017) Deep convolutional neural networks for image classification: a comprehensive review. Neural Comput 29:2352–2449
39. Rumelhart DE, Hinton GE, Williams RJ (1986) Learning representations by back-propagating errors. Nature 323:533–536
40. Schmidhuber J (2015) Deep learning in neural networks: an overview. Neural Networks 61:85–117
41. Simard PY, Steinkraus D, Platt JC (2003) Best practices for convolutional neural networks applied to visual document analysis. In: Icdar
42. Simonyan K, Zisserman A (2014) Very deep convolutional networks for large-scale image recognition. arXiv Prepr arXiv14091556
43. Szegedy C, Liu W, Jia Y, Sermanet P, Reed S, Anguelov D, Erhan D, Vanhoucke V, Rabinovich A (2015) Going deeper with convolutions. In: Proceedings of the IEEE conference on computer vision and pattern recognition, pp 1–9
44. Tan C, Sun F, Kong T, Zhang W, Yang C, Liu C (2018) A survey on deep transfer learning. In: International conference on artificial neural networks. Springer, pp 270–279
45. Veres M, Moussa M (2019) Deep learning for intelligent transportation systems: a survey of emerging trends. IEEE Trans Intell Transp Syst
46. Voulodimos A, Doulamis N, Doulamis A, Protopapadakis E (2018) Deep learning for computer vision: a brief review. Comput Intell Neurosci
47. Wang Y, Zhang D, Liu Y, Dai B, Lee LH (2019) Enhancing transportation systems via deep learning: a survey. Transp Res Part C Emerg Technol 99:144–163
48. Xu B, Wang N, Chen T, Li M (2015) Empirical evaluation of rectified activations in convolutional network. arXiv Prepr arXiv150500853
49. Yosinski J, Clune J, Bengio Y, Lipson H (2014) How transferable are features in deep neural networks? In: Advances in neural information processing systems. pp 3320–3328
50. Young T, Hazarika D, Poria S, Cambria E (2018) Recent trends in deep learning based natural language processing. IEEE Comput Intell Mag 13:55–75
51. Zeiler MD, Fergus R (2013) Stochastic pooling for regularization of deep convolutional neural networks. arXiv Prepr arXiv13013557
52. Zeiler MD, Fergus R (2014) Visualizing and understanding convolutional networks. In: Processing of European conference on computer vision, Zurich, Switzerland, 5–12 Sept 2014
53. Zhao Q, Griffin LD (2016) Suppressing the unusual: towards robust cnns using symmetric activation functions. arXiv Prepr arXiv160305145
54. Zhao Z-Q, Zheng P, Xu S, Wu X (2019) Object detection with deep learning: a review. IEEE Trans Neural Networks Learn Syst 30:3212–3232
55. Zhuang F, Cheng X, Luo P, Pan SJ, He Q (2015) Supervised representation learning: Transfer learning with deep autoencoders. In: Twenty-Fourth international joint conference on artificial intelligence

Leveraging CNN Deep Learning Model for Smart Parking

Guruvareddiyur Rangaraju Karpagam, Abishek Ganapathy, Aadhavan Chellamuthu Kavin Raj, Saravanan Manigandan, J. R. Neeraj Julian, and S. Raaja Vignesh

Abstract The automated car parking system is a system which helps the people to park their vehicles without any confusion in a mall or hospital or theatre or any parking layout. This automated car parking system takes the footage of the parking layout which is given as input and is made standard for the model. It is used to find empty and filled parking slots. Then with that information, it directs the user or the person who comes inside the parking to the empty slot or gives a message that the parking is full. The system is trained using Deep learning with different images of parking slots with empty cars and parking slots with the car filled. The objective is to leverage the CNN Deep Learning model for Smart Parking. The main problem in car parking is the improper management of land resources, which leads to a great shortage in parking space causing chaos in our daily lives. Parking is one of the biggest challenges that we need to tackle in the years to come. Using smart parking, there is proper utilization of the parking space and makes the experience pleasant.

G. R. Karpagam (✉) · A. Ganapathy · A. C. Kavin Raj · S. Manigandan · J. R. Neeraj Julian · S. Raaja Vignesh
Department of Computer Science and Engineering, PSG College of Technology, Coimbatore, Tamil Nadu, India
e-mail: grk.cse@psgtech.ac.in

A. Ganapathy
e-mail: abishekganapathy15592@gmail.com

A. C. Kavin Raj
e-mail: kavinraj.c1999@gmail.com

S. Manigandan
e-mail: manisaravanan11@gmail.com

J. R. Neeraj Julian
e-mail: neerajjulian@gmail.com

S. Raaja Vignesh
e-mail: iamraajavigneshs@gmail.com

K. R. Ahmed et al. (eds.), *Deep Learning and Big Data for Intelligent Transportation*, Studies in Computational Intelligence 945,
https://doi.org/10.1007/978-3-030-65661-4_8

Keywords Smart parking system · Artificial intelligence · Deep learning · CNN · DNN · Region of interest · Precision · Recall · Confusion matrix · ROC curve · AUC curve

1 Motivation

Car parking is one of the most prominent problems that the world faces. Following the rapid increase in car ownership, many cities find it difficult to manage their parking areas with a huge imbalance between parking supply and demand [1]. This shortage may be mainly attributed to faulty use of land and miscalculation of parking requirements. Now the main problem in a parking lot is finding if there is any free space or not. When a driver enters the parking lot first he/she needs to know if there are any free spaces and then the latter to know which spot is free for a hassle-free experience. Therefore, offering drivers with relevant information about the parking lot while entering the parking lot has become an important issue in today's world. We have to develop a system by keeping a check on a few things. The main issue is we should achieve the objective with no requirement for additional resources.

The objective is to find all the parking spaces and find if a parking space is available or occupied using the Deep Learning Model. Efforts have been taken to design a system that uses image processing to detect the existence of the car and also provides information such as the number of available parking spaces and the location of such free spaces. The system captures images using a webcam and processes the image to count the available parking space. The development of this system will use techniques of image processing that will be implemented in each phase of the methodology. This system gives information about the number of available parking spaces. Counting available parking space using image processing to solve the problem that the driver faces, at a low cost. There are enough perquisites of using a camera-based system as compared to the other existing systems. Firstly, there is no requirement for additional infrastructure, provided the facility is operated on CCTV. Secondly, camera-based systems are highly precise [2–4].

2 Related Work

Some of the existing works in the area of parking slot detection are as follows:

Chiu et al. (2004) suggested a method for counting the vehicles at the checkpoint from which the number of available parking spaces can be counted [5]. The counting in this method would be performed by the installation of the induction loop sensors under the road surface. The main disadvantage in this method is that the sensors are very difficult to maintain, in case of malfunction and are very difficult to install it causing damage to the roads. Moreover, the exact locations of the free parking area cannot be determined because the counting method would not able to give detailed information as it just records the number of vehicles passing the checkpoints.

Park et al. (2008) proposed the use of ultrasonic sensors mounted on the cars to search for a free parking space [4]. This method suffers from a drawback that the sensors are easily affected by weather conditions like rain, temperature, snow and fast air breeze.

Boda et al. (2007) also presented a method based on wireless sensor nodes [6]. This method was less costly and it uses the wireless sensor nodes implemented at the critical places like the lane turns, entrance and exit positions of the parking lot. The total number of cars in the parking area can be determined by the difference between incoming and outgoing cars. This method also suffers from drawbacks similar to those in the previous two methods such as damage due to external sources and the inability to detect the exact location of vacant parking spaces.

Some of the existing works in the area of parking slot detection based on image classification are:

Bin et al. (2009) proposed a vision based method where the whole parking area available for parking can be examined through a camera and the data obtained is then processed and the result generated will determine the exact number and location of vacant parking spaces [7]. A major drawback of this method is that the accuracy of the output is highly dependent on the position of the camera.

Fabian (2008) also proposed an unsupervised vision based system for parking space occupancy detection [8]. The proposed system has low complexity in computation and needs fewer image frames per minute. He claims that the major problem in image detection is the occlusions and shadows, unsupervised learning requires more advanced clustering algorithms. The vision based parking space detection systems are mostly affected by weather and lighting conditions like the falling of rain drops on the lens of the camera during heavy rainfall, low and high lighting conditions.

Yusnita et al. (2012) presented a method in which a brown colour round patch was drawn in each parking space manually [7]. When the system is initialized it looks for the rounded shape in each space if a patch is detected that particular space is considered as free and will be displayed to the driver. When the patches are blocked by objects (vehicles) then the system assumes the particular spaces are filled by vehicles. The system was good enough for managing the parking lot, however it does not work well in heavy rainfall and snow, another drawback being persons or objects in parking lots which might be identified as vehicles.

True (2007) proposed an efficient parking space detection by using the combinations of colour histogram and vehicle features detection [9]. This method is efficient compared to the previous methodologies but is not accurate enough. Considering parking spaces with a little number of parking lots, even a small error percentage may mean a large mistake.

2.1 Lessons Learnt

Based on the literature survey this section outlines the lessons learnt is as follows:

- This smart parking requires a CNN deep learning model which outperforms other conventional methods using hand crafted features for the detection of parking occupancy in terms of accuracy and robustness.
- Using a CNN model would be better compared to the other available deep learning models owing to the factor that convolutional neural networks work better with image processing comparatively.

3 Context of the Work

The context of the work is described in Fig. 1.

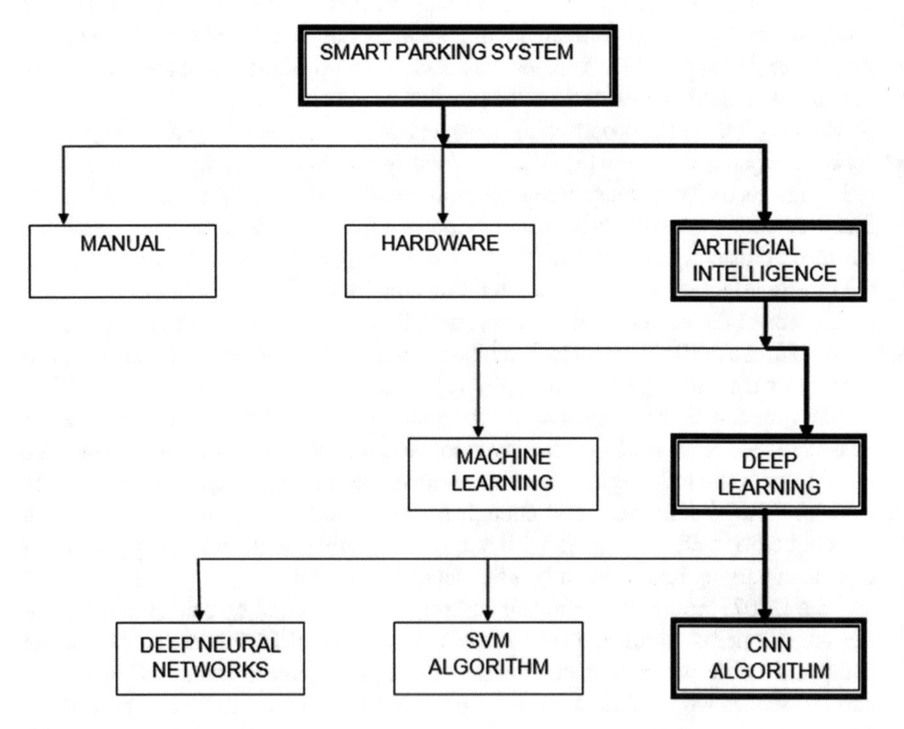

Fig. 1 Context of the work

3.1 *Machine Learning*

Machine Learning is the study of algorithms and statistical models that computer systems use to perform a specific task without using explicit instructions, relying on patterns and inference instead. It is seen as a subset of artificial intelligence. Machine learning algorithms build a mathematical model based on sample data, known as "training data". In order to make predictions or decisions without being explicitly programmed to perform the task. Machine learning algorithms are used in a wide variety of applications, such as email filtering and computer vision, where it is difficult or infeasible to develop a conventional algorithm for effectively performing the task.

Machine learning tasks are classified into several broad categories. In supervised learning, the algorithm builds a mathematical model from a set of data that contains both the inputs and the desired outputs. For example, if the task were to determine whether an image contained a certain object, the training data for a supervised learning algorithm would include images with and without that object (the input), and each image would have a label (the output) designating whether it contained the object. In special cases, the input may be only partially available or restricted to special feedback.

Classification algorithms and regression algorithms are types of supervised learning. Classification algorithms are used when the outputs are restricted to a limited set of values. For a classification algorithm that filters emails, the input would be an incoming email, and the output would be the name of the folder in which to file the email. For an algorithm that identifies spam emails, the output would be the prediction of either "spam" or "not spam", represented by the Boolean values true and false. Regression algorithms are named for their continuous outputs, meaning they may have any value within a range. Examples of a continuous value are the temperature, length, or price of an object.

In unsupervised learning, the algorithm builds a mathematical model from a set of data that contains only inputs and no desired output labels. Unsupervised learning algorithms are used to find structure in the data, like grouping or clustering of data points. Unsupervised learning can discover patterns in the data and can group the inputs into categories, as in feature learning.

Reinforcement learning is an area of Machine Learning. It is about taking suitable action to maximize reward in a particular situation. It is employed by various software and machines to find the best possible behavior or path it should take in a specific situation. Reinforcement learning differs from the supervised learning in a way that in supervised learning the training data has the answer key with it so the model is trained with the correct answer itself whereas in reinforcement learning, there is no answer but the reinforcement agent decides what to do to perform the given task. In the absence of a training dataset, it is bound to learn from its experience [10].

3.2 Deep Learning

Deep learning is an artificial intelligence function that imitates the workings of the human brain in processing data and creating patterns for use in decision making. Deep learning is a subset of artificial intelligence (AI) that has networks capable of learning unsupervised data that is unstructured or unlabelled. Also known as deep neural learning or deep neural network [1, 11].

Deep learning has evolved hand-in-hand with the digital era, which has brought about an explosion of data in all forms and from every region of the world. This data, known simply as big data, is drawn from sources like social media, internet search engines, e-commerce platforms, and online cinemas, among others. This enormous amount of data is readily accessible and can be shared through fine tech applications like cloud computing.

However, the data, which normally is unstructured, is so vast that it could take decades for humans to comprehend it and extract relevant information. Companies realize the incredible potential that can result from unraveling this wealth of information and are increasingly adapting to AI systems for automated support.

3.3 Comparison of Various Models for Image Classification

DNN is a model that can be used for classification. We can successfully train a simple neural network to perform regression and classification, but a DNN may not perform well with images.

Convolution layers have proved to be very successful in tasks involving images such as image classification, object identification and face recognition. They allow parameter sharing which results in a very optimized network compared to using Dense layers. Convolutional Neural Networks achieve high performance compared to alternative methods in the field of classification, due to the strong and rich features from which they can learn from the large data through deep architecture.

Support Vector Machine algorithm works great for image classification but will require us to design our own kernel or treat our data before using them.

Logistic regression and decision trees can also be used for image classification. But they give significant errors at prediction time.

Comparing all the above models we see that CNN is good for image classification because:

(1) CNN compares the image piece by piece. So, CNN gets better seeing similarity than whole image matching schemes.
(2) Since CNN is a kind of deep learning neural network, there is transfer learning that takes place, and hence it learns more and results in output with less error.
(3) CNN can be thought of as automatic feature extractors from the image.
(4) CNN uses adjacent pixel information to effectively down sampling of the image first by convolution and then uses a prediction layer at the end.

(5) CNN performs feature extraction and gives better accuracy.

Comparison of models in the context of Smart parking system using Deep learning Models

Model	Description	Pros	Cons
Deep neural networks	A deep neural network is a neural network with a certain level of complexity, a neural network with more than two layers. Deep neural networks use sophisticated mathematical modeling to process data in complex ways	It is comparatively simple	DNN does not perform well with images
Support vector machine	Support-vector machines are supervised learning models with associated learning algorithms that analyze data used for classification and regression analysis Support-vector machines construct a hyperplane or set of hyperplanes in a high- or infinite-dimensional space, which can be used for classification, regression or other tasks like outliers detection	Experimental results show that SVMs achieve significantly higher search accuracy than traditional query refinement schemes after just three to four rounds of relevant feedback	It involves a considerable amount of preprocessing. It requires us to design our own kernel or treat our data before using them
Logistic regression	Logistic regression is a supervised learning classification algorithm used to predict the probability of a target variable. The nature of the target or dependent variable is dichotomous, which means there would be only two possible classes	Logistic regression is easier to implement, interpret, and very efficient to train. It is very fast at classifying unknown records	It constructs linear boundaries and gives significant errors at prediction time

(continued)

(continued)

Model	Description	Pros	Cons
Convolutional neural networks	The name convolutional neural network indicates that the network employs a mathematical operation called convolution. Convolution is a specialized kind of linear operation. Convolutional networks are simply neural networks that use convolution in place of general matrix multiplication in at least one of their layers	CNN compares the image piece by piece. So CNN gets better seeing similarity than whole image matching schemes and provides better accuracy through feature extraction	CNN does not encode the position and orientation of the object and lacks the ability to be spatially invariant to the input data

3.4 Mathematical Model

CNN is a combination of Biology, Arts and Mathematics. The Neural Networks are in the core of all Deep Learning Algorithms. They try to replicate the brain and its workings process [3].

The CNN has two main components:

- **The convolution layers** that extracts features from the input and,
- **The fully connected (dense) layers** that uses data from convolution layer to generate output.

There are also two important processes involved in the training of any neural network:

- **Forward Propagation:** Receive input data, process the information, and generate output
- **Backward Propagation:** Calculate error and updates the various parameters of the network

Forward Propagation:

A convolutional neural network identifies the picture's shape and edges by comparing its pixel values.

Convolution is often represented mathematically with an asterisk * sign. If we have an input image represented as X and a filter represented with f, then the expression would be:

$$Z = X * f$$

where * indicates element-wise multiplication of X and f

Dimension of image = (n, n)

Dimension of filter = (f, f)

Dimension of output will be ((n − f + 1), (n − f + 1)).

The fully connected layer takes in only 1D matrix as input. So, the previous layer inputs are converted into 1D matrix for further processing. The fully connected layer performs two operations on the incoming data—**a linear transformation and a non-linear transformation**.

The **linear transformation** is performed by:

$$Z = W^T \cdot X + b$$

where X is the input, W is weight, and b (called bias) is a constant. Also, the W in this case will be a matrix of (randomly initialized) numbers so that the hidden layers are not identical to each other.

The **non-linear transformation** is added to the data to capture additional complex relationships in the data. The component that does the non-linear transformation is called the Activation Function.

For this work, the activation function used is ReLU (Rectified Linear Unit) for the hidden units and sigmoid function in the output unit. The mathematical expression for this is:

$$\begin{aligned} A &= \max(0, Z) \quad \text{for the hidden layers} \\ A &= 1/1 + e^{-Z} \quad \text{for the output layer (For binary classification)} \end{aligned}$$

The activation function is added to all the layers in the neural network.

Backward Propagation:

During the forward propagation process, we randomly initialized the weights, biases and filters. These values are treated as parameters from the convolutional neural network algorithm. In the backward propagation process, the model tries to update the parameters such that the overall predictions are more accurate.

To start backpropagation, we find the cost function J by using the gradient descent algorithm. We try to find the parameter values as those values that tried to reduce this cost function J. The general equation to update parameters are:

$$new_{parameter} = old_{\ parameter} - learning_{\ rate} * gradient\_of\_parameters$$

where the $\text{learning}_{\text{rate}}$ is a hyperparameter that controls the change in the $\text{new}_{\text{parameter}}$ based on the $\text{old}_{\text{parameter}}$

To update weight, do:

$$W_{new} = W_{old} - learning_{\ rate} * \frac{dE}{dW}$$

where

$$\frac{dE}{dW} = \frac{dE}{dO} \cdot \frac{dO}{dZ} \cdot \frac{dZ}{dW}$$

To update bias(b) do:

$$b_{new} = b_{old} - learning_{rate} * \frac{dE}{db}$$

where

$$\frac{dE}{db} = \frac{dE}{dO} \cdot \frac{dO}{dZ} \cdot \frac{dZ}{db}$$

For the convolution layer, we had the filter matrix as our parameter. To update the filter matrix, do:

$$f = f - learning_{rate} * \frac{dE}{df}$$

where

$$\frac{dE}{df} = \frac{dE}{dO} \cdot \frac{dO}{dZ} \cdot \frac{dZ}{dA} \cdot \frac{dA}{dZ} \cdot \frac{dZ}{df}$$

4 Conceptual Architecture

The intent of conceptual architecture is to direct attention at an appropriate decomposition of the system without delving into the details of interface specification. Key constructs are identified, including significant architectural elements such as components and relationships among them, as well as architectural mechanisms

By focusing on key constructs and abstractions rather than a proliferation of technical details, conceptual architecture provides a useful vehicle for communicating the architecture to non-technical audiences, such as management, marketing, and in some cases users. It is also the starting point for Logical Architecture, which elaborates the component specifications and architectural mechanisms to make the architecture precise and actionable. The conceptual architecture for the project is given in Fig. 2:

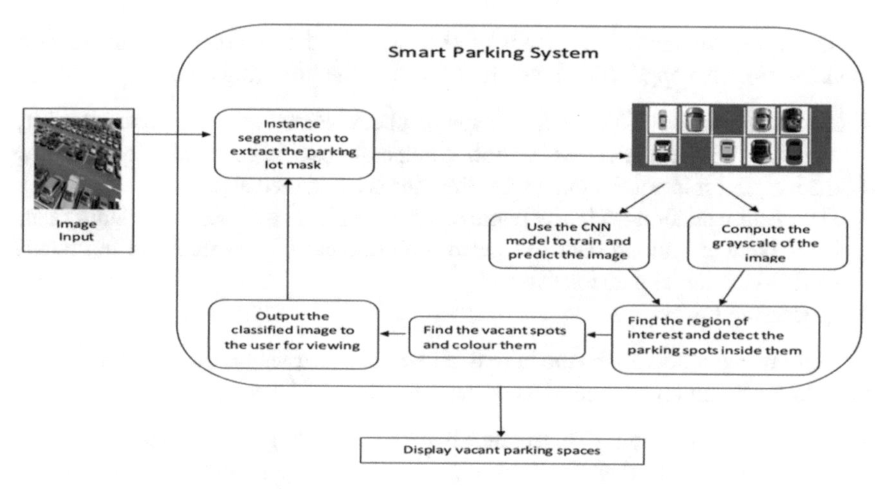

Fig. 2 Conceptual architecture

Application layer		**Communication manager**
Image capture interface	Exception handler	
Pre-processing layer		
Image checking	Data augmentation	
Processing layer		
GrayScale conversion	Image segmentation	
Hough transformation	Region of interest manipulator	
Lane determinator	Slot determinator	
Physical layer		
Storage manager	Image database	
Result database	Maintenance manager	

The conceptual design of the project is modelled as a 4-layer architecture. They are as follows:

- Application Layer

This layer contains all the modules related to the real-life application of the project. It consists of the following modules:

- Image capture Interface - This is the interface that allows us to capture the image of the parking lot via CCTV cameras which further helps us to find the empty parking slots.
- Exception Handler - This module helps us to report any kind of errors or exceptions the program might face while executing.

- Pre-processing Layer

This layer contains all the modules that helps in the preprocessing of the data obtained from the application layer. It consists of the following modules:

- Image Checking—This module helps us check whether the image captured by the CCTV camera is correct or not. It helps the user in accepting or discarding the image if it is not the image that the user is concerned about.
- Data Augmentation—This step allows us to artificially create new training data from existing training data. We perform horizontal flip, vertical and horizontal shifts to the initial data received.
- Processing Layer

This layer contains all the modules that helps in processing the data obtained from the preprocessing layer. It consists of the following modules:

- GrayScale conversion—This module helps in converting the processed image into grayscale which further allows us to process the image to find the empty slots.
- Image segmentation—This module helps us to segment the image into smaller parts. It helps in detecting the various edges present in the image.
- Region of Interest Manipulator—This module helps in manipulating the image by showing us the part of the image that interests us by removing all the unnecessary regions in the image.
- Hough transformation—This module performs the hough transformation on the image that helps in isolating features of the region of interest within an image.
- Lane determinator—This module helps in drawing lines indicating the various parking lanes present in the image.
- Slot determinator—This module helps in identifying the parking slots and drawing a box (and colouring it) around the empty parking slots.

- Physical Layer

This layer contains all the modules that helps in the storage and maintenance of the images obtained from the processing layer. It consists of the following modules:

- Storage manager—This manager helps in storing images into various databases. It helps to store images into the Image and Result database.
- Image database—This is the image database that stores all the images taken by the CCTV camera at any point of time. It also stores the timestamp of all images indicating the time and date of when the photo was captured.
- Result database—This is the result database that stores the final processed image for further checks or clarification by the user, if needed.
- Maintenance manager—This manager helps in notifying the user when the maintenance is required for the system.

One module common to all the above mentioned layers is the **Communication Manager**. This manager helps the program communicate easily between all the layers, to perform its tasks.

4.1 Algorithm

INPUT: An image that is to be classified
OUTPUT: A classified image showing the vacant parking spots

PROCESS:

1. Start
2. Initialize X and Z to be two empty lists
3. Get all the car images required to detect if the object in the spot is a car or not
4. For each image:
   - Store the processed image into X
   - Store the image's label into Z
5. Create another variable Y that stores the Z labels as 0 and 1
6. Create x_train,x_test,y_train,y_test from the set of X and Y values
7. Create a CNN model
8. To prevent overfitting of the CNN model use data augmentation
9. Training is done through forward propagation and backpropagation for many, many epochs. This repeats until we have a well-defined neural network with trained weights and feature detectors.
10. Perform prediction on the model by giving test inputs(x_test) and evaluate it
11. Initialize prop_class=[]
12. Initialize mis_class=[]
13. For i=0 to len(y_test):
    - if(y_test[i]==prediction[i]):
      - prop_class.append(i)
14. For i=0 to len(y_test):
    - if(y_test[i]!=prediction[i]):
      - mis_class.append(i)
15. Display the correctly classified and misclassified images
16. Get the image of the required parking lot
17. Convert the image into grayscale image

18. Find the region of interest in the parking lot
    - Get the image's shape
    - Using trial and error find out 6 polygonal points
    - Use this polygonal points to mask the region of interest
19. Use hough lines to detect straight lines in the image, and mark them
20. Initialize cnt_empty = 0
21. Initialize all_spots = 0
22. Determine parking lanes
23. For each parking lanes:
    - Find the parking spots
    - For each parking spot:
        - Use CNN model to predict if there is a car present in the spot or not
        - If (label=="Empty"):
            - Colour the spot green
            - cnt_empty+=1
        - all_spots+=1
24. Display the predicted image

- Assign_label() and make_data():-
  – The Assign_label() software component is used to assign labels to all the training data available so that training on these labeled data can be performed
  – The make_data() software component is used to split all the data into training and testing data
- Make_model():-
  – It's used to build the CNN software component along with applying various steps to ensure that this model doesn't overfit the data
- Train_model():-
  – This software component is used to train the model based on the data that is available
- Evaluate_model():-
  – Used to evaluate the accuracy and the loss of the CNN model built
- Prediction():-

  - Used to predict the corresponding labels for the test data input to the CNN model

- Convert_gray_Scale():-
  - The main parking lot image is converted into grayscale so that the detection of vacant parking spots would be easier
- Select_region():-
  - This software component is used to select the region of interest in which we want to predict the vacant parking space
- Hough_line():-
  - This software component is used to identify the edges present in the image so that an outline of each parking spot can be detected
- Identify_lanes() and identify_spots():-
  - This software component helps to identify the various lanes and spots present in the parking lot image
- Predict_image():-
  - This software component is used to display the final image showing all the vacant spots in the entire parking lot

The pictorial flow of the Smart parking system is explained in the Appendix (A6–A18).

5 Assessment

This section gives the description of the dataset used in this work, the various metrics used to evaluate the work along with the metric graphs.

The total data used for this work is: 381 images

The total Training data used for this work is: 285 images

The total Test data used for this work is: 96 images

True positive is data points which are identified as positive which are actually positive. False positive is data points which are identified as positive but which are actually negative. True negative is Data points which are identified as negative which are actually negative. False negative is data points which are identified as negative but which are actually positive.

The values of TP, TN, FP, FN for work is tabulated as follows:

Binary classification	Values
True positive	62
False positive	2
True negative	32
False negative	0

5.1 *Recall and Precision Metrics*

Recall:

It is an ability of a classification model that identifies all relevant instances.
The recall value of the software component is 1.0000.

Precision:

It is an ability of a classification model that identifies only relevant instances.
The precision value of our software component is 0.96875.

F1 Score:

F1 score combines recall and precision by harmonic mean.
The F1 score of the software component is 0.984127.

5.2 *Visualising Recall and Precision*

Confusion Matrix:

It shows the predicted and actual table from the classification problem. The confusion matrix for the smart parking system is shown in Fig. 3.

Receiver Operating Characteristic (ROC) Curve:

The ROC curve is required to plot the true positive rate (TPR) and false positive rate (FPR) with respect to the model threshold for separating the positives. The ROC curve is shown in Fig. 4.

Fig. 3 Confusion matrix

```
Confusion Matrix:

array([[32,  2],
       [ 0, 62]])
```

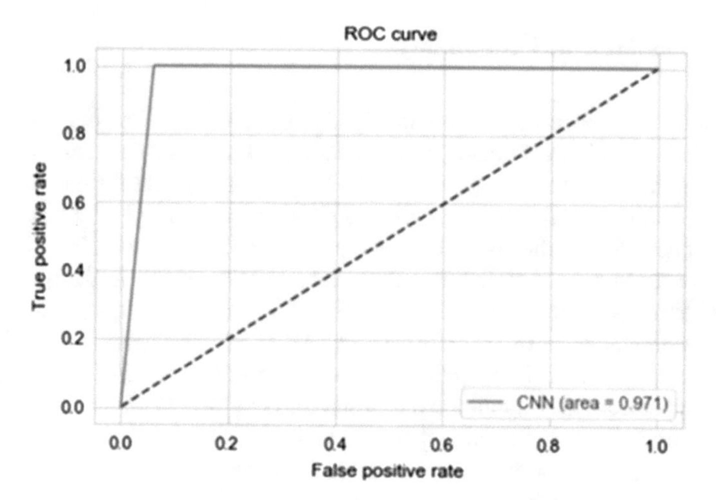

Fig. 4 ROC curve

Area Under the Curve (AUC) Curve:

The AUC is used to calculate the performance of a model by finding the area under a ROC curve [3]. The AUC curve is shown in Fig. 5.

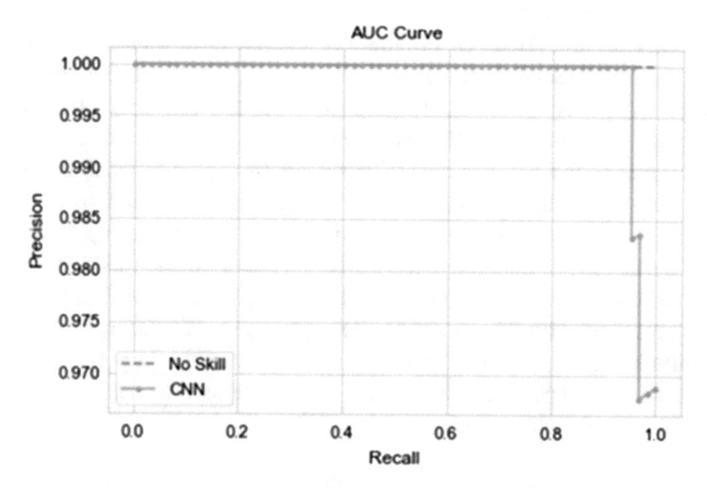

Fig. 5 AUC curve

6 Discussion and Conclusion

In the modern day world people are indulged in transportation and thus face a lot of problems. The problem of finding a vacant parking slot is considered as our problem statement. The requirement explained as the Motivation explains the problem people face while attempting to park their car in a parking space. They might search for a parking space without the knowledge of available parking spaces. Thus, the project helps the user to gain knowledge of the available parking slot in the given space. The problem has been solved in the past by many other techniques, contrastingly we have attempted to solve the problem using Artificial Intelligence(Deep learning). The context explains Artificial Intelligence in depth so that the user understands the concept and how it can be applied to this problem. The plot explains the different algorithms available in Deep Learning that can be used to solve the problem. Comparing the various algorithms CNN is chosen as it is the best for image processing. The photo of the parking space given as input is processed by the CNN algorithm and the user gets to know whether a parking slot is available in the space. The image given as input is converted into grayscale images and broken into small pixels to determine whether the slot is vacant or occupied. The efficiency of the model is derived from the outcome of binary classification and recalling of the precision metrics. The model developed works perfectly for the given input and any change in layout leads to a change in determination of Region of Interest which varies for every parking layout. Thus, the model can be used by the user to determine the availability of parking slots in the given parking layout.

Deep Learning and Artificial Intelligence are the most blooming topics in the current Computer Science world. However, in order to have performance rather than having a good algorithm training the dataset is also critically important. This deep learning based Parking Slot detection Model is used as an efficient application to provide services to the customer/user. This Model will continuously be developed to provide services to the user based on real image obtained through the CCTV or live camera feed thus allowing the model to be used for any parking layout helping the user/customer more efficiently. The CNN algorithm can be improvised by giving input of a more efficient parking layout and by using a precise dataset which is used for efficient classification. The model can be improvised in the future by transforming it into a mobile application so that users can know whether there is a vacant space remotely without entering into the layout. This prevents the user from going back from the layout in the absence of a parking slot.

Acknowledgements We would thank Impiger Technologies Pvt. Ltd for their support and help throughout our project.

Appendix 1

See Figs. 6, 7, 8, 9, 10, 11, 12, 13, 14, 15, 16, 17, 18 and 19.

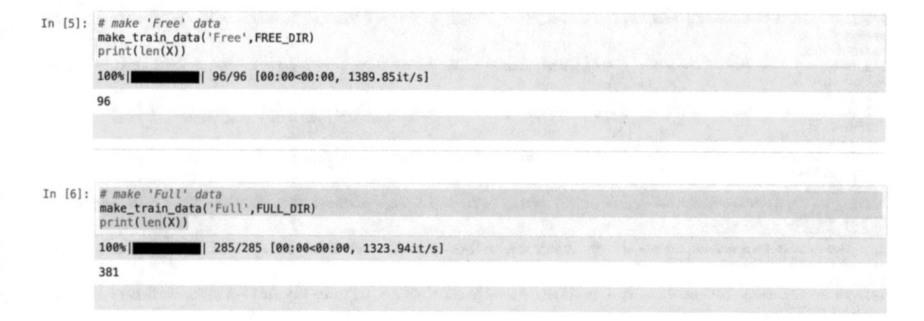

```
In [5]: # make 'Free' data
        make_train_data('Free',FREE_DIR)
        print(len(X))

        100%|██████████| 96/96 [00:00<00:00, 1389.85it/s]

        96

In [6]: # make 'Full' data
        make_train_data('Full',FULL_DIR)
        print(len(X))

        100%|██████████| 285/285 [00:00<00:00, 1323.94it/s]

        381
```

Fig. 6 Loading the dataset

```
In [15]: model.summary()

Model: "sequential_1"
_________________________________________________________________
Layer (type)                 Output Shape              Param #
=================================================================
conv2d_1 (Conv2D)            (None, 48, 48, 32)        2432
_________________________________________________________________
max_pooling2d_1 (MaxPooling2 (None, 24, 24, 32)        0
_________________________________________________________________
conv2d_2 (Conv2D)            (None, 24, 24, 64)        18496
_________________________________________________________________
max_pooling2d_2 (MaxPooling2 (None, 12, 12, 64)        0
_________________________________________________________________
conv2d_3 (Conv2D)            (None, 12, 12, 96)        55392
_________________________________________________________________
max_pooling2d_3 (MaxPooling2 (None, 6, 6, 96)          0
_________________________________________________________________
conv2d_4 (Conv2D)            (None, 6, 6, 96)          83040
_________________________________________________________________
max_pooling2d_4 (MaxPooling2 (None, 3, 3, 96)          0
_________________________________________________________________
flatten_1 (Flatten)          (None, 864)               0
_________________________________________________________________
dense_1 (Dense)              (None, 512)               442880
_________________________________________________________________
activation_1 (Activation)    (None, 512)               0
_________________________________________________________________
dense_2 (Dense)              (None, 2)                 1026
=================================================================
Total params: 603,266
Trainable params: 603,266
Non-trainable params: 0
_________________________________________________________________
```

Fig. 7 Summary of a model

```
Epoch 38/50
2/2 [==============================] - 0s 241ms/step - loss: 0.1534 - accuracy: 0.9618 - val_loss: 0.0800 - val_acc
uracy: 0.9583
Epoch 39/50
2/2 [==============================] - 0s 231ms/step - loss: 0.1069 - accuracy: 0.9554 - val_loss: 0.1383 - val_acc
uracy: 0.9479
Epoch 40/50
2/2 [==============================] - 1s 301ms/step - loss: 0.1033 - accuracy: 0.9727 - val_loss: 0.0889 - val_acc
uracy: 0.9688
Epoch 41/50
2/2 [==============================] - 1s 269ms/step - loss: 0.0297 - accuracy: 1.0000 - val_loss: 0.0736 - val_acc
uracy: 0.9792
Epoch 42/50
2/2 [==============================] - 0s 245ms/step - loss: 0.0880 - accuracy: 0.9809 - val_loss: 0.0617 - val_acc
uracy: 0.9792
Epoch 43/50
2/2 [==============================] - 0s 249ms/step - loss: 0.0525 - accuracy: 0.9682 - val_loss: 0.0950 - val_acc
uracy: 0.9688
Epoch 44/50
2/2 [==============================] - 1s 267ms/step - loss: 0.1789 - accuracy: 0.9745 - val_loss: 0.0619 - val_acc
uracy: 0.9792
Epoch 45/50
2/2 [==============================] - 1s 369ms/step - loss: 0.0539 - accuracy: 0.9805 - val_loss: 0.0472 - val_acc
uracy: 0.9688
Epoch 46/50
2/2 [==============================] - 0s 245ms/step - loss: 0.0721 - accuracy: 0.9682 - val_loss: 0.0742 - val_acc
uracy: 0.9688
Epoch 47/50
2/2 [==============================] - 1s 363ms/step - loss: 0.0532 - accuracy: 0.9883 - val_loss: 0.1239 - val_acc
uracy: 0.9583
Epoch 48/50
2/2 [==============================] - 1s 254ms/step - loss: 0.0380 - accuracy: 0.9873 - val_loss: 0.0558 - val_acc
uracy: 0.9792
Epoch 49/50
2/2 [==============================] - 0s 236ms/step - loss: 0.0434 - accuracy: 0.9809 - val_loss: 0.0529 - val_acc
uracy: 0.9688
Epoch 50/50
2/2 [==============================] - 0s 247ms/step - loss: 0.0916 - accuracy: 0.9618 - val_loss: 0.0601 - val_acc
uracy: 0.9792
```

Fig. 8 Training the model

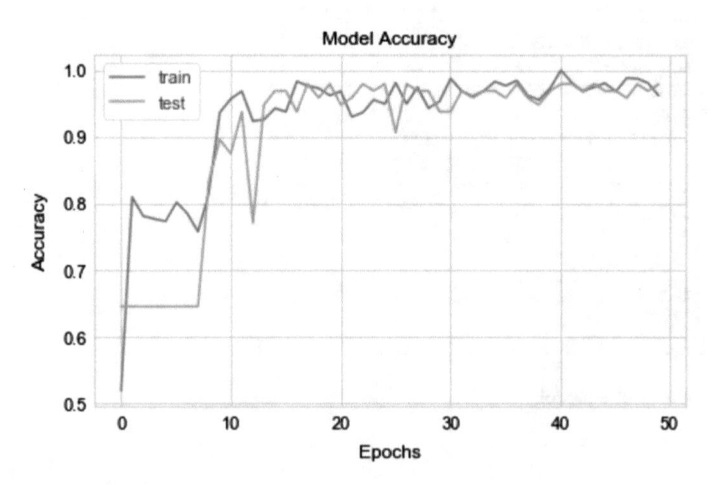

Fig. 9 Model accuracy

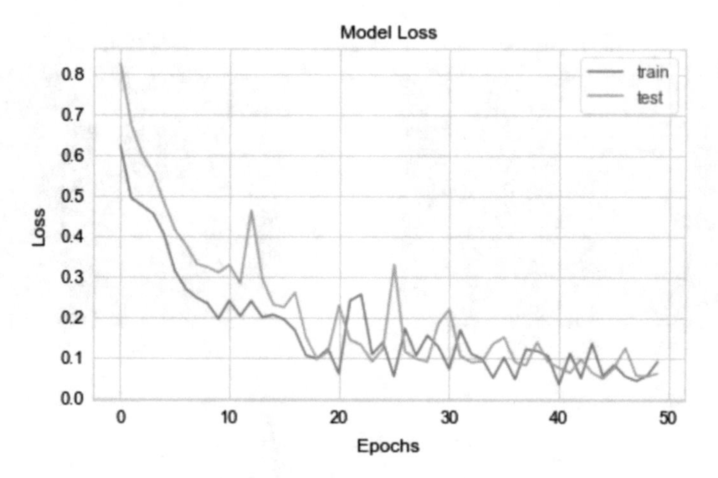

Fig. 10 Model loss

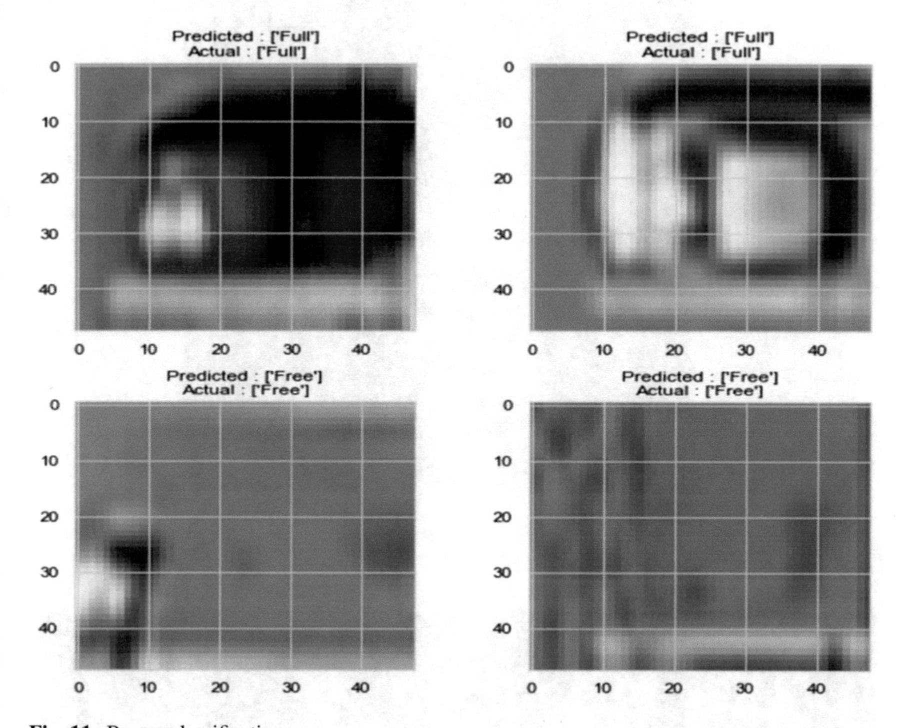

Fig. 11 Proper classification

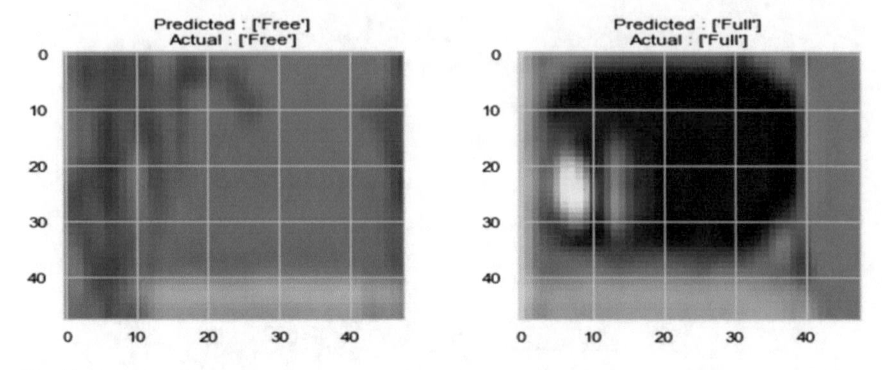

Fig. 11 (continued)

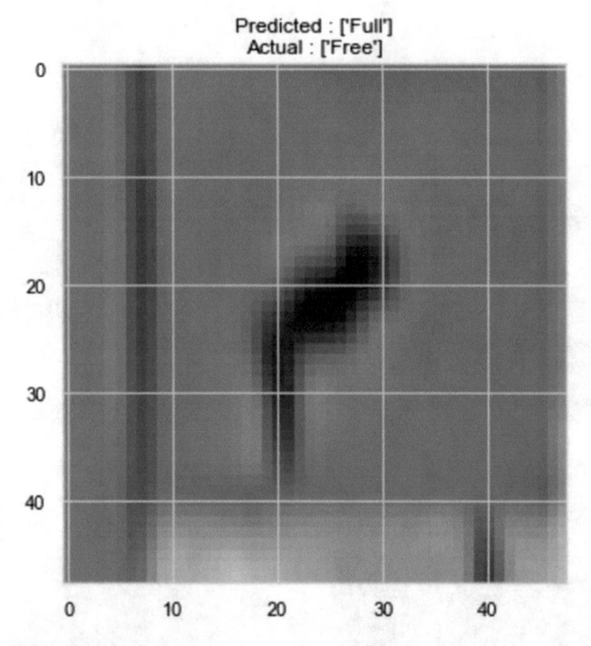

Fig. 12 Improper classification

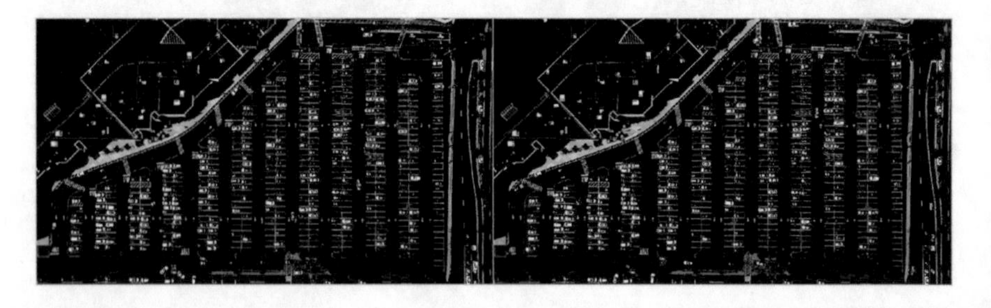

Fig. 13 Grayscale conversion

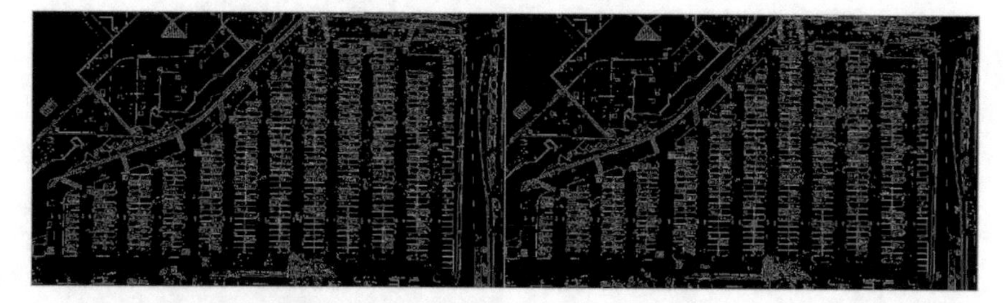

Fig. 14 Determining the edges of grayscale conversion

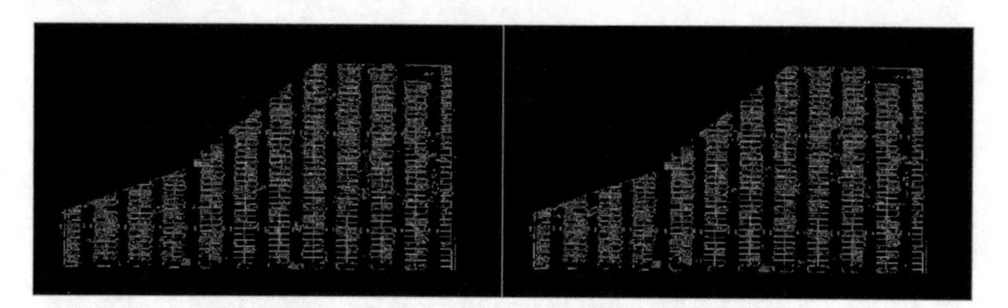

Fig. 15 Region of interest

```
No lines detected:  569
No lines detected:  588
```

Fig. 16 Determination of outline of parking slots

```
Num Parking Lanes:  12
Num Parking Lanes:  12
```

Fig. 17 Determination of parking lanes

Fig. 18 Final layout of slots and lanes

Fig. 19 Figure of the final output

References

1. Acharya D, Khoshelham K, Winter S (2017) Real-time detection and tracking of pedestrians in cctv images using a deep convolutional neural network. In: Proceeding of the 4th annual conference of research@locate, vol 1913
2. https://towardsdatascience.com/beyond-accuracy-precision-and-recall-3da06bea9f6c
3. https://www.analyticsvidhya.com/blog/2020/02/mathematics-behind-convolutional-neural-network/
4. Park W-J, Kim B-S, Seo D-E, Kim D-S (2008) Parking space detection using ultrasonic sensor in parking assistance system. In: Intelligent vehicles symposium. IEEE
5. Chiu M-Y, Depommier R, Spindler T (2004) An embedded real-time vision system for 24-hour indoor/outdoor car-counting applications. Pattern Recogn
6. Boda VK, Nasipuri A, Howitt I (2007) Design considerations for a wireless sensor network for locating parking spaces. In: SoutheastCon
7. Yusnita R, Fariza N, Norazwinawati B (2012) Intelligent parking space detection system based on image processing. Int J Innov Manag Technol 3(3)
8. Fabian T (2008) An algorithm for parking lot occupation detection. In: Computer information systems and industrial management applications
9. True N (2007) Vacant parking space detection in static images. University of California, Projects inVision & Learning
10. Mitchell T (1997) Machine learning. McGraw Hill
11. Bin Z, Dalin J, Fang W, Tingting W (2009) A design of parking space detector based on video image. Electron Measure Instrum

Estimating Crowd Size for Public Place Surveillance Using Deep Learning

Deevesh Chaudhary, Sunil Kumar, and Vijaypal Singh Dhaka

Abstract With the overwhelming speed of population outbursts throughout the world, it raises a security concern over public places like supermarkets, offices, banks, political rallies, religious events etc. The risk of any abnormal behavior or any security concern arises with the number of people in the crowd. Most of the public places are overcrowded and require crowd count monitoring to avoid any mishappening. It is impossible to appoint personnel at different locations to count people over there. The role of CCTV cameras is pertinent as far as remote monitoring of crowds over public places is concerned, but it is a cumbersome job for a person to count people in a crowd only by monitoring multiple videos of different locations at a time. Also, with the help of CCTV footage it is not an easy task to manage crowds, specifically if it is a dense crowd. Automated crowd count that can estimate the total number of people in a crowd image is needed for the hour. In this chapter, we have reviewed crowd count methods using state of art deep learning models for automated crowd count and their performance analysis on major crowd counting datasets.

1 Introduction

Public surveillance is continuous monitoring, analysis, and interpretation of real time videos through CCTV cameras installed at various locations. The purpose of public surveillance is to remotely monitor the activities of people by watching videos on screen captured through CCTV cameras. Public video surveillance has become one

D. Chaudhary
Department of Information Technology, Manipal University Jaipur, Jaipur, Rajasthan, India
e-mail: deevesh.choudhary@jaipur.manipal.edu

S. Kumar (✉) · V. S. Dhaka
Department of Computer and Communication Engineering, Manipal University Jaipur, Jaipur, Rajasthan, India
e-mail: skvasistha@gmail.com

V. S. Dhaka
e-mail: vijaypalsingh.dhaka@jaipur.manipal.edu

K. R. Ahmed et al. (eds.), *Deep Learning and Big Data for Intelligent Transportation*, Studies in Computational Intelligence 945,
https://doi.org/10.1007/978-3-030-65661-4_9

of the major topics of interest due to its increasing demand in various applications like crowd behavior analysis, crowd count, public security, urban planning etc. In recent years automated crowd count has attracted many researchers from the computer vision field. Research in this area has been widely explored to help human beings in real time scenarios like detection of any abnormal activity, counting number of people in the crowd, tracking human movement etc.

In this chapter, we are going to discuss the most advanced technology i.e. deep learning methods for automated public surveillance and its various objectives. Although, researchers from the computer vision field have been working hard for automation in the estimation of people's count within the crowd at public places, Convolutional Neural Network (CNN) seems to provide promising results in this field. CNN is a sub-branch of neural networks that can extract and learn salient features from an image to estimate the number of people.

1.1 Why Do We Need Automated Crowd Count?

Let us illustrate a situation to understand why there is a need for an automated crowd count. Imagine a conference of data scientists conducted in a university that has many sessions related to different topics and interests. We want to analyze the effectiveness of sessions based on the number of people attending these sessions and based on that it will be easier to manage the next conference gathering. For this, we need several people to attend each session. It can be done by manually counting the number of people in each session, but it requires manpower and time. Another approach is to take photographs of each session and build a computational model that takes this photograph as input and estimates the count of people in the photograph as output. In this way, it would be quite easier to estimate the count of people from the image for which minimum manpower is required and we get results in less time. Similarly, there are plenty of events where automated crowd count seems purposeful.

- Counting number of people attending political rallies.
- Counting the number of people at religious events in temples.
- Counting number of terrorists at border.
- Monitoring high congestion traffic.
- Managing exit plans in shopping malls and cinema halls.

Broadly there are plenty of use cases of automated crowd count from commercial to defense purpose.

1.2 Crowd Count: Definition and Background

Crowd count refers to counting the number of objects in a still image or video. As shown in Fig. 1, objects may be people, cells, cars, any living or non-living being.

Fig. 1 Top row (L to R): people and traffic count, bottom row (L to R): cell and fruit count

Our focus will be on estimating the number of people in dense crowd images. Due to its wide variety of applications in various fields such as medical, defense, disaster management, public surveillance, health care, urban planning etc. there is a need to shift from traditional methods of manual counting through CCTV to more advanced and automated methods of crowd count that give more accurate and precise results.

Technically, people count from an image can be done with digital image processing, machine learning and deep learning as categorized in Fig. 2.

We are going to discuss most advanced technology i.e. deep learning methods (CNN/ConvNet) to estimate crowd count but before that we will have some overview of machine learning methods for crowd count. Traditional state of art methods for crowd count includes count by detection [1–3], count by regression [4–6], count by density estimation [7, 8], count by clustering [9–11].

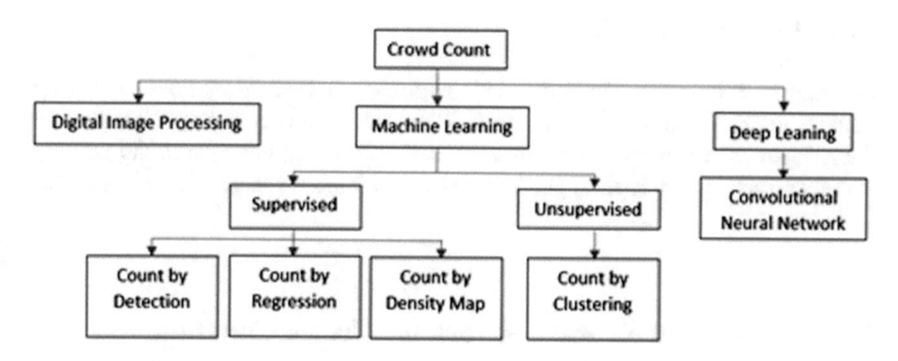

Fig. 2 Automated crowd count techniques

1.2.1 Count by Detection

In this approach, a sliding window detector is used to identify either monolithic (full body detection) or part based (specific body parts like head, shoulder) features from the image. A well-trained classifier is required to extract low level features such as HOG [12], SIFT, Edgelet [13], background subtraction [14] etc. Count by detection works well in sparse crowd image but for dense crowd image accuracy drops drastically.

1.2.2 Count by Regression

In this approach a regression model is used to generate mapping between estimated and actual count. Initially, low level features are extracted from the image and then regression modelling (linear, ridge, gaussian etc.) is used to estimate count. In this approach, the image is divided into small patches of common interest like people moving in common direction or low-level features (shape, edge, texture) of each patch is extracted. A regression function that maps feature vectors to crowd size is applied to estimate the crowd count. Different types of regression models can be applied such as linear, ridge, gaussian, lasso regression etc.

1.2.3 Count by Density Estimation

Unlike count by detection approach, this approach performs the detection and localization of complete human body or certain parts of human body. Occlusion and clutter problems are also tackled by obtaining the density map of crowd image. Estimated count is obtained by integrating the density map. This approach works well for dense crowds as compared to other approaches of crowd count.

1.2.4 Count by Clustering

This approach based on similarity patterns among individual motions. Moreover, visual features of individuals are also considered while clustering the similar features. These features are grouped into different categories based upon similarity patterns. However, this approach gives good results in continuous image frames rather than static images.

2 Convolutional Neural Network for Crowd Counting

Estimating count with the help of low-level hand-crafted features using above approaches give good results to some extent. Researchers have found that deep learning approach i.e. by using CNN on crowd related images is able to extract more robust and deep features from image. Using different conv layers and fine-tuning hyperparameters, exceptional results are achieved compared to traditional approaches. Though using CNN is computationally complex but performance of CNN for feature extraction is better as compared to other approaches.

CNN is a computationally complex approach, but proves to be one of the influential innovations in the computer science field. Researchers from computer vision found that the practical application of Convolutional Neural Network (CNN or ConvNets) provides promising results as far as image processing is concerned. To estimate crowd count from any image we provide an image dataset to CNN model to train the model. Input image passes through basic layers of convnet i.e. convolutional, ReLu, pooling and fully connected (FC) layers. Depending on complexity, images may go through multiple combinations of convolution and ReLu layers having variable filter size and hyperparameters. The image dataset for crowd count consists of various images as shown in Fig. 3. CNN model after preprocessing an image gives an estimated count of people. A general model of how CNN works on image specifically for count estimation is illustrated in Fig. 3.

Researchers from the computer vision field have proposed a number of CNN models (Multicolumn CNN [15], Switching CNN [16], Cascaded CNN [17], CSRNet [18], Deep CNN [19]. Few CNN models can estimate count even better than humans.

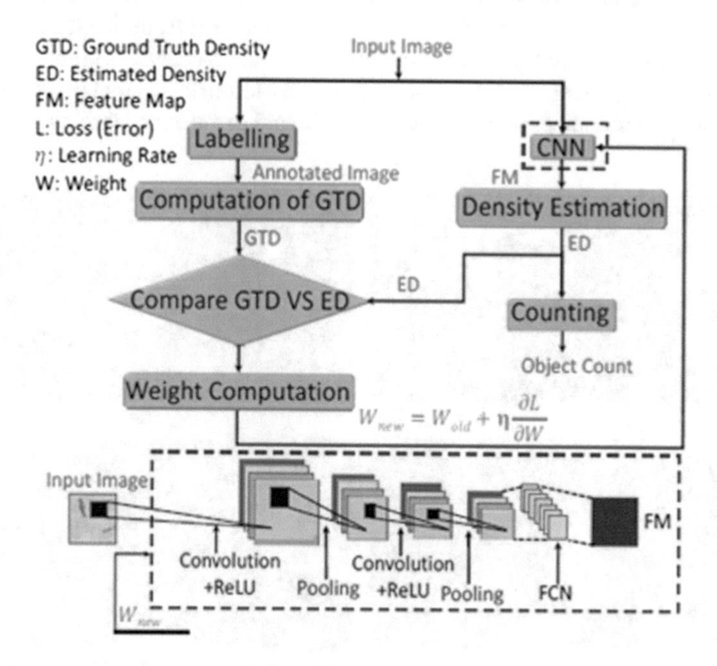

Fig. 3 Methodology of CNN on crowd image [16]

2.1 Why CNN?

There is reason to shift from the traditional methods of crowd counting to deep learning methods (CNN) due to following reasons.

- Counting number of people from an image encounter some challenges such as occlusion(people overlapping each other), high clutter(people moving in different directions), varying light illumination(daylight image or night light image), different scale and inconsistent perspective(different camera angle images) etc. which causes prediction error and low accuracy. Due to these challenges, it is difficult to count people in dense crowds using traditional methods mentioned above. Illustration of these challenges are shown in Fig. 4.
- All the above-mentioned approaches focused upon low-level hand-crafted features. These features can be identified and extracted successfully if the crowd is sparse. For dense crowd counting, estimated count changes drastically for these methods.
- Due to varying application of automated crowd counting for the public to the military and its importance in computer vision, crowd counting has become a complicated scientific problem to be solved. There is a need for a technique that can provide more accurate estimates of count in densely populated areas.
- Another advantage of using CNN over handcrafted feature extraction approaches is that it automatically extracts most important features from the dataset without any human intervention. The feature extraction task for CNN is computationally efficient and gives precise results as compared to other techniques. Also, CNN uses a smaller number of parameters as compared to machine learning algorithms

Fig. 4 Challenges **a** Occlusion. **b** Different camera angle. **c** Varying illumination. **d** Cluttering

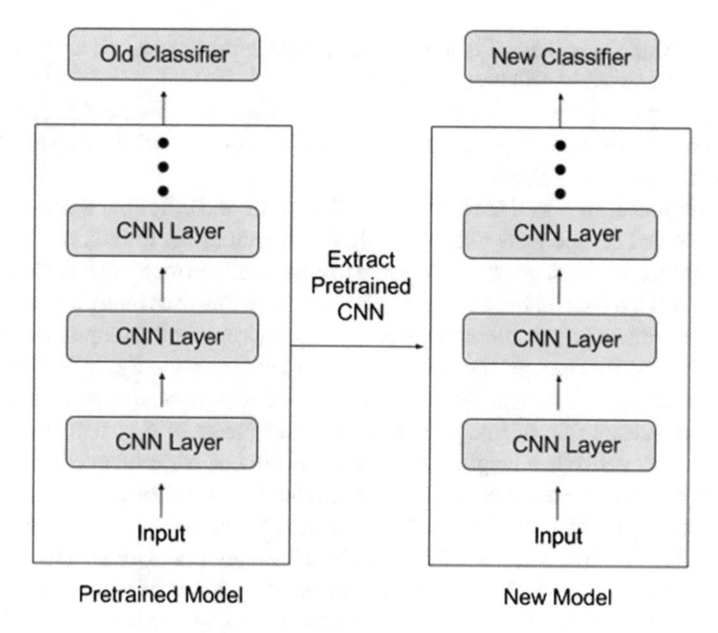

Fig. 5 Transfer learning with pre trained model

such as regression, classification, SVM etc. Various operations in CNN i.e. convolutional, pooling, flatten enables parameter sharing and thus make the model more robust.

- Pre-trained CNN models can be used for transfer learning also for related problems in computer vision. In that way we can reduce training time and can also improve efficiency of models. We can reuse or tune by changing hyperparameters and configuration of preexisting models. Examples of a few pre-trained models are Oxford VGG, Google Inception, Microsoft ResNet, AlexNet etc. Figure 5 illustrates how transfer learning works.

The basic idea of transfer learning is to leverage the pre-trained deep learning models layers with existing weights and not to change weights during training while using the model for new tasks.

2.2 CNN Architecture

CNN is arguably one of the most popular deep learning architectures for image/video related classification tasks. The sudden surge of interest of researchers in CNN models is due to its effectiveness and on the go model behavior. The very first CNN working model for image related tasks AlexNet was developed in 2012 with 8 layers.

Since then AlexNet gained so much popularity that within 3 years, 8 layers AlexNet has been progressed to 152 layers ResNet.

All CNN models have somewhat similar architectures. A researcher can change the number of layers and hyperparameters as per his requirement and dataset. A basic architecture of CNN is shown in Fig. 6.

Major layers in any CNN are convolutional, pooling and fully connected. There are other terms and components related to CNN architecture i.e. filter/kernel, padding, stride, softmax. In this section we are going to discuss all the major layers and components used in CNN. CNN image related problem takes image as input, processes it by passing through various combinations of layers and provides the output. Before dive into CNN, let us first understand how an image is taken as input by CNN. Computers do not see images as human eyes. Computers see images as arrays of pixels that depend upon resolution of image. For example an image of $5 \times 5 \times 3$ resolution represents $h \times w \times d$ (h = height, w = width, d = dimension/number of channels) of image. Resolution is measured in terms of number of pixels horizontal x number of pixels vertical x number of channels. Thus image of resolution $5 \times 5 \times 3$ is an array of matrices of size $5 \times 5 \times 3 = 75$ pixels of RGB values(Red, Green, Blue) whereas an image of resolution $5 \times 5 \times 1 = 25$ represents an array of matrix of grayscale image. Figure 7 illustrates how an image is read by computer in terms of pixel values in the form of matrix arrays. The pixel values in the array represent intensity of color.

By using suitable instruction as per language, an input image is first converted into a pixel array. For e.g. in python we use numpy asarray() class to convert PIL images into numpy arrays. This array gets processed through several CNN layers for feature extraction and the model gets trained by updating the weights using backpropagation.

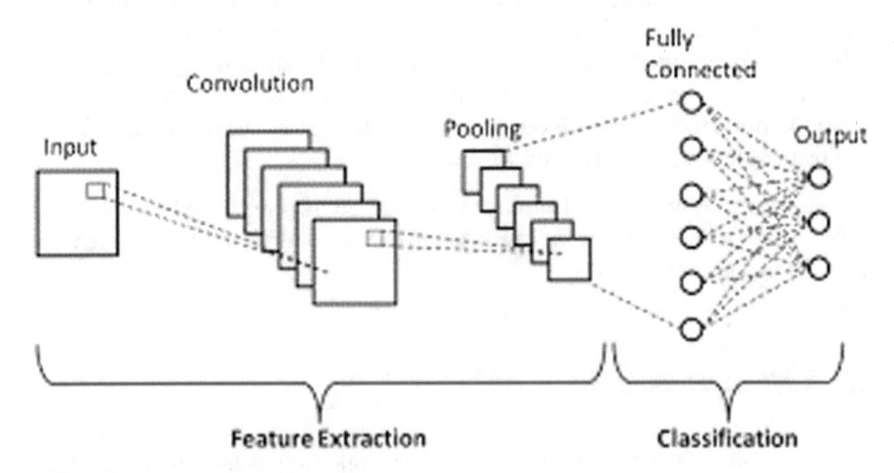

Fig. 6 Basic architecture of CNN

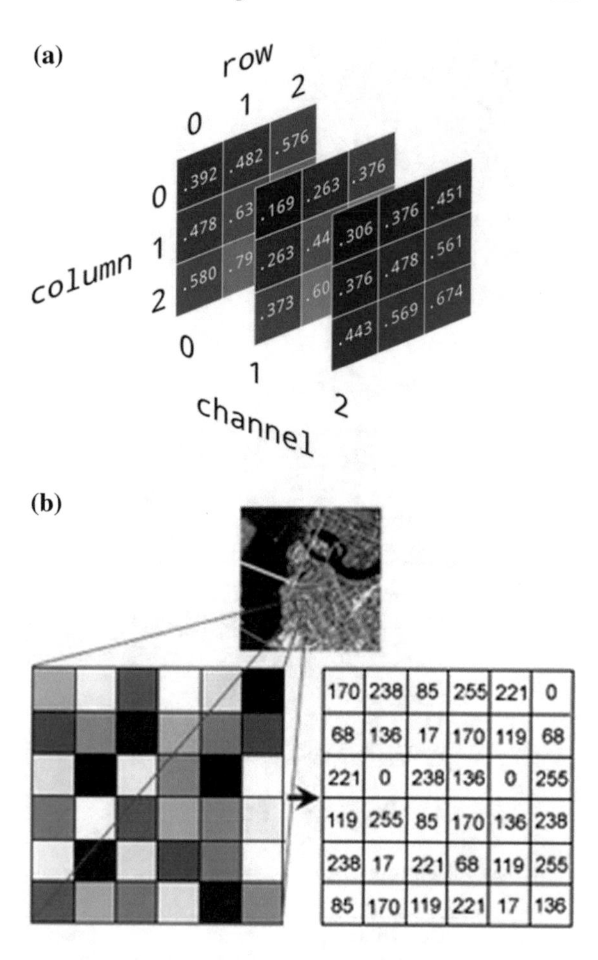

Fig. 7 **a** Colored image pixel values in 3 RGB channels. **b** Grayscale image pixel in single channel

2.2.1 Convolution Layer

This is the main building block and very first layer of any CNN architecture. Convolution is simply a mathematical function like multiplication that tends to merge information from two different sets to single. It takes two inputs i.e. image matrix and kernel/filter. An image of dimension $h \times w \times d$ and a filter of dimension $fh \times fw \times fd$ will output a volume of dimension $(h - fh + 1) \times (w - fw + 1)$ named as feature map or convolved feature. Convolution operation is illustrated in Fig. 8. Convolution operation is performed by sliding a filter over the input matrix step by step. At each step element wise matrix multiplication is performed and sums the result. This sum will go into a new matrix called a feature map. Figure 9 shows popular filter used to extract image features.

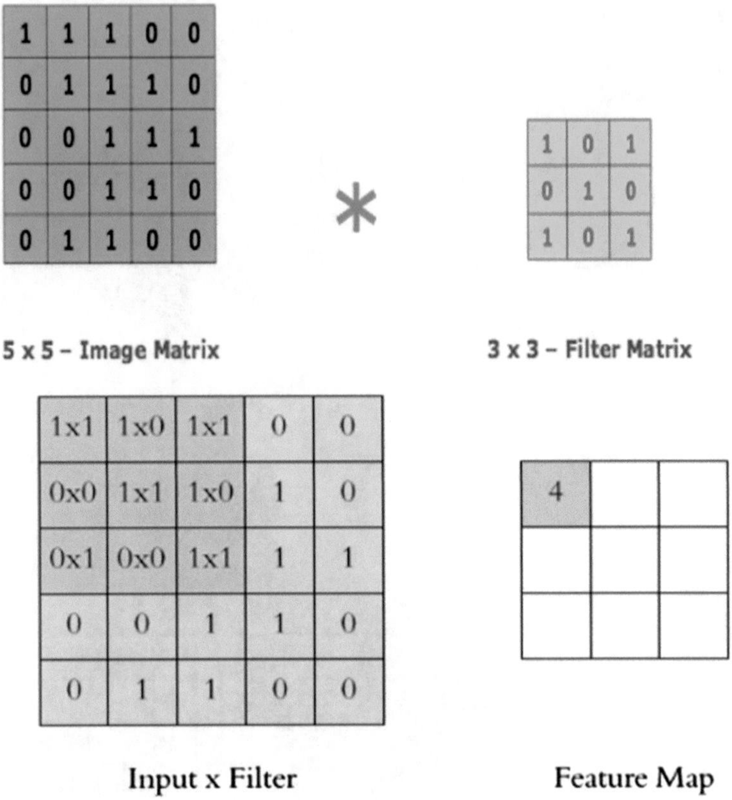

1	1	1	0	0
0	1	1	1	0
0	0	1	1	1
0	0	1	1	0
0	1	1	0	0

1	0	1
0	1	0
1	0	1

5 x 5 – Image Matrix

3 x 3 – Filter Matrix

1x1	1x0	1x1	0	0
0x0	1x1	1x0	1	0
0x1	0x0	1x1	1	1
0	0	1	1	0
0	1	1	0	0

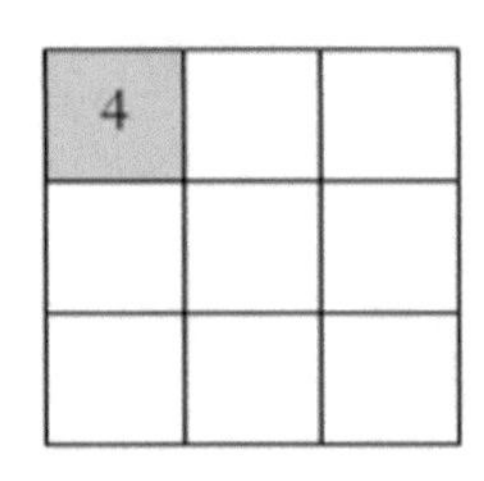

Input x Filter

Feature Map

Fig. 8 Convolution operation

Operation	Filter
Identity	$\begin{bmatrix} 0 & 0 & 0 \\ 0 & 1 & 0 \\ 0 & 0 & 0 \end{bmatrix}$
Edge Detection	$\begin{bmatrix} 1 & 0 & -1 \\ 0 & 0 & 0 \\ -1 & 0 & 1 \end{bmatrix}$ $\begin{bmatrix} 0 & 1 & 0 \\ 1 & -4 & 1 \\ 0 & 1 & 0 \end{bmatrix}$ $\begin{bmatrix} -1 & -1 & -1 \\ -1 & 8 & -1 \\ -1 & -1 & -1 \end{bmatrix}$
Sharpen	$\begin{bmatrix} 0 & -1 & 0 \\ -1 & 5 & -1 \\ 0 & -1 & 0 \end{bmatrix}$
Normalized	$\frac{1}{9}\begin{bmatrix} 1 & 1 & 1 \\ 1 & 1 & 1 \\ 1 & 1 & 1 \end{bmatrix}$
Gaussian Blur	$\frac{1}{16}\begin{bmatrix} 1 & 2 & 1 \\ 2 & 4 & 2 \\ 1 & 2 & 1 \end{bmatrix}$

Fig. 9 Popular filters used for convolution

Fig. 10 ReLU layer

2.2.2 ReLU(Rectified Linear Unit)

It is an activation function that is applied after convolution operation. It produces a linear output for positive values and output zero for negative values as shown in Fig. 10. The values in the feature map matrix are converted to rectified feature map by application of ReLU function. The activation function is applied to implement nonlinearity in CNN and achieve higher accuracy. Various activation functions can be used in CNN like tanh, sigmoid but ReLU is the most frequently used activation function. Few advantages of using ReLU over other activation functions are computational efficiency, representational sparsity, linear behavior, and easy to train model.

2.2.3 Stride and Padding

Stride means how far the filter will move at each step over the input image matrix. By default, stride value is 1. Figure 11 illustrates how filters will move over the input matrix for different stride values.

It must be noted that larger the stride size, smaller will be the output size. For higher stride size, the size of the feature map becomes small. To maintain same dimensionality, we use padding i.e. to surround the input matrix with zeros as shown in Fig. 12. In other words, padding is used to preserve the dimensionality of the feature map otherwise it will shrink after each layer.

The area around the input image is padded with zeros. After padding, the size of the feature map is the same as the original image dimension except for the number of channels or depth. To get the same dimensionality of feature map and input with stride = 1, the padding size should be (k − 1)/2 where k is kernel size. In general formula to calculate output size with respect to filter size, stride and padding is:

$$O = \frac{W - K + 2P}{S} + 1$$

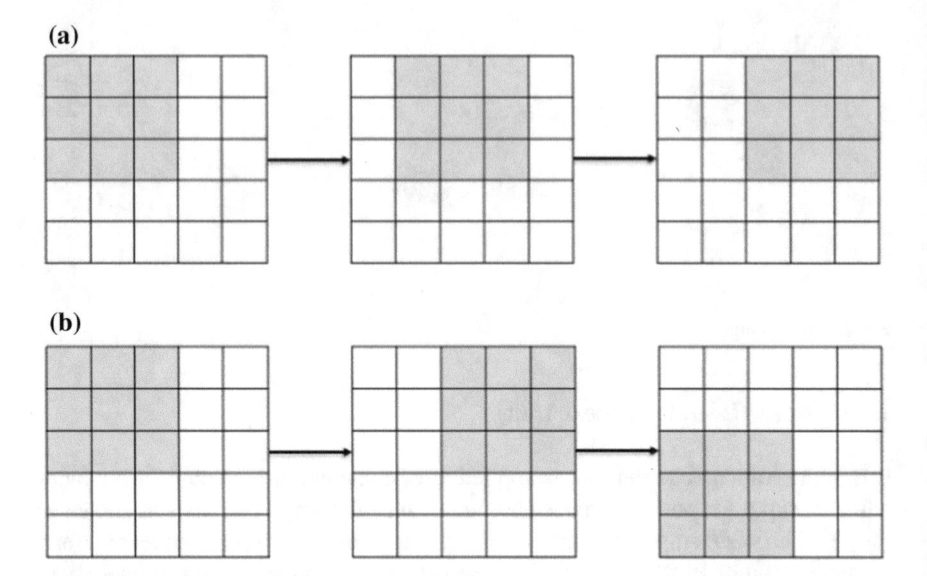

Fig. 11 **a** Stride = 1. **b** Stride = 2

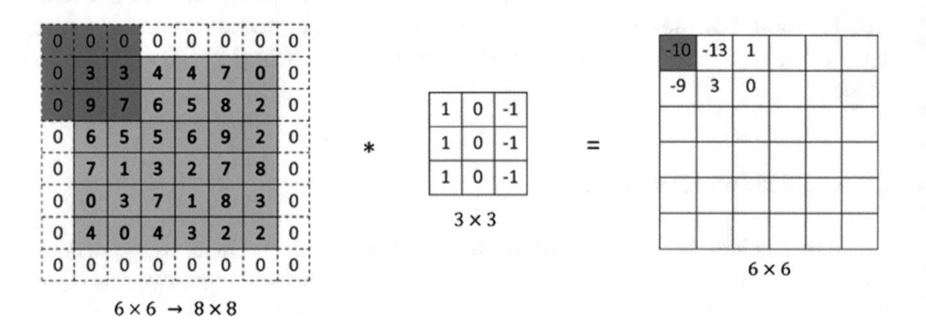

Fig. 12 Padding operation

where O is the output height/length, W is the input height/length, K is the filter size, P is the padding, and S is the stride. For example, if we take S = 1, P = 2 with W = 200 and K = 5 and use 40 filters, then the output size will be 200 × 200 × 40. On the other hand if we use S = 1, P = 1, then the output size would be 198 × 198 × 40.

2.2.4 Pooling

Also known as sub sampling or down sampling, pooling layer helps in reducing number of parameters or dimensionality of large image feature map but retaining the important features simultaneously. This layer down sample each rectified feature

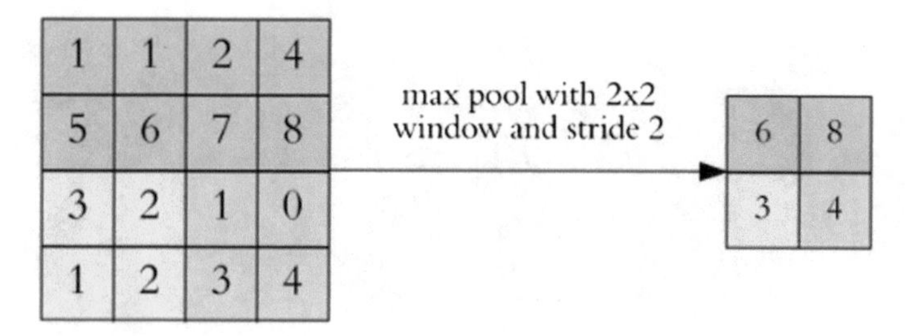

Fig. 13 Example of max pooling

map individually, thus reducing the dimension of the feature map, keeping the depth unchanged. This down sampling helps in fast training of models but also overcomes overfitting of models. Pooling is of three types: max pooling, sum pooling, and average pooling. Most common type is max pooling that takes max value in the feature map pooling window with specified size. Similarly, in the case of average pooling it takes the average of all values in the pooling window. Figure 13 illustrates pooling operation with 2×2 window size and stride $= 2$. Each color in the figure denotes an individual window.

It must be noted that with this window size and stride, feature map size becomes half but still retaining the important features. For example, if the feature map has dimension $32 \times 32 \times 8$ and using window size 2×2 with stride $= 2$, then after pooling operation we obtain a window of size $16 \times 16 \times 8$. This is the main use of pooling, down sampling the rectified feature map while keeping important features intact. By down sampling, we reduce the number of weights to a considerable amount thus reducing the training time of the model. In most of the CNN architectures, pooling operation is done with window size 2×2, stride $= 2$ and padding $= 0$. However, we can change these parameters to obtain desired results.

2.2.5 Fully Connected (FC) Layer

The output of (Conv + ReLU + Pooling) operation is 3D vector volumes, therefore the output of the final pooling layer is flattened to obtain an 1D vector array which acts as input to FC layer. After passing through a fully connected layer, the final layer uses sigmoid or softmax function to get probabilities of output classes. Y Figure 14 illustrates a typical CNN architecture.

All the popular CNN architectures viz. AlexNet, VGGNet, ResNet, GoogleLeNet use the same basic architecture, however with different hyper parameters like size of filters, type of pooling, number of hidden layers, stride size, padding type, and size etc.

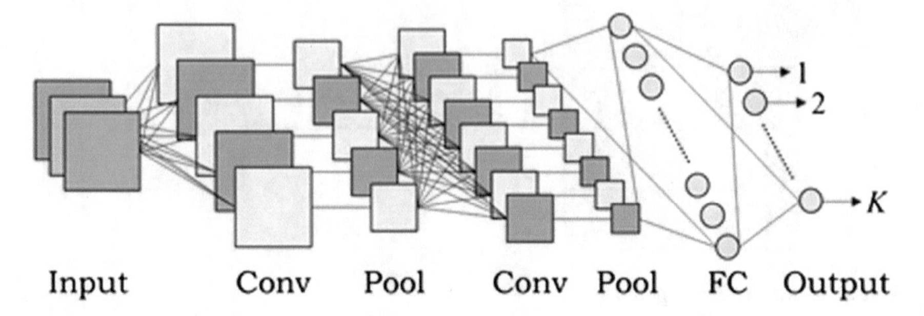

Fig. 14 Typical CNN architecture

3 State of the Art CNN Models for Crowd Counting

In this section we are going to discuss the most popular CNN models for crowd counting viz. Multicolumn CNN [15], Switching CNN [16], Cascaded CNN [17] and CSRNet [18].

3.1 Multicolumn CNN

Zhang et al. [15] have proposed a multi column convolutional neural network that maps images into its density map. They developed a method to count the number of people from a single image having arbitrary density and perspective. The dataset considered in this work consists of images as shown in Fig. 15. It can be observed that the image consists of a heavily occluded crowd scene and there is large variation in scale of people/head size due to perspective effect. The traditional detection-based methods will not work on these kinds of images. Also, foreground segmentation of this kind of image dataset is a cumbersome job. To overcome these issues,

Fig. 15 Shanghai tech dataset image

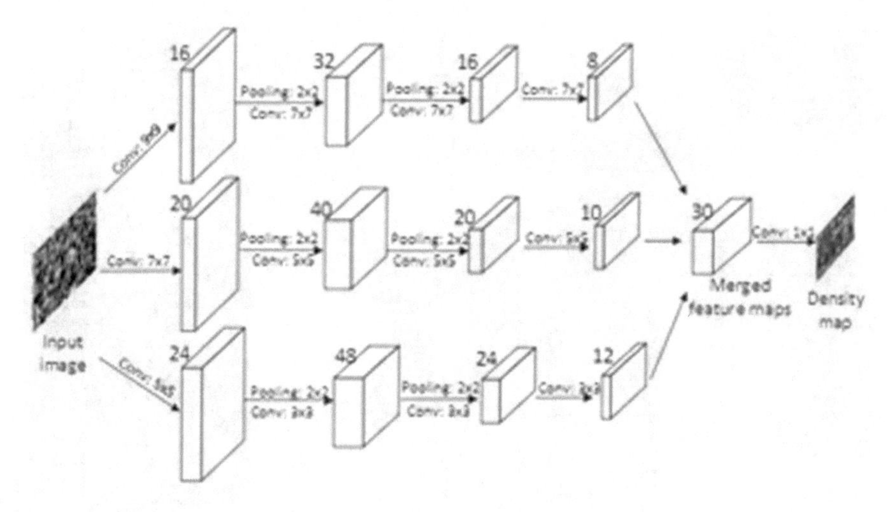

Fig. 16 MCNN architecture [15]

authors have suggested multicolumn neural network architecture that consists of three columns having different filter sizes corresponding to receptive fields of different sizes(small, medium and large) to cope up with large variation in size of people/head.

The learning of CNN model is done through density maps of dataset obtained by geometry adaptive kernel. Structure of the multicolumn CNN is shown in Fig. 16. Three parallel CNN are used corresponding to different filter size receptive fields i.e. $9 \times 9, 7 \times 7, 5 \times 5$. Less number of larger size filters are used to reduce complexity.

A very basic combination of conv-pool-conv-pool is used in all three CNN's. Max pooling is applied for each 2×2 region with RELU activation function. The output feature map of three CNNs are merged and mapped to the final density map by using filter size of 1×1.

3.2 Switching CNN

Sam et al. [16] proposed an approach that is robust to large scale and perspective variations of people in crowd scene scenarios. In this approach, crowd scene image is divided into patches and a switch classifier is used to relay each patch on an independent CNN regressor (R1, R2 and R3) based upon density of crowd. These regressors are designed to have different receptive fields corresponding to different people's size. R1 filter size is 9×9 and has a small receptive field to capture high level abstraction in patches such as faces. This regressor is suitable for high density patches. R2 consists of filter size 7×7 and R3 consists of filter size of 5×5 for different scale variations. Depending on distortion, perspective and density level of patch, patches are relayed to the best CNN regressor with the help of switch classifier. The quality of regressors is based upon crowd count prediction quality during CNN training.

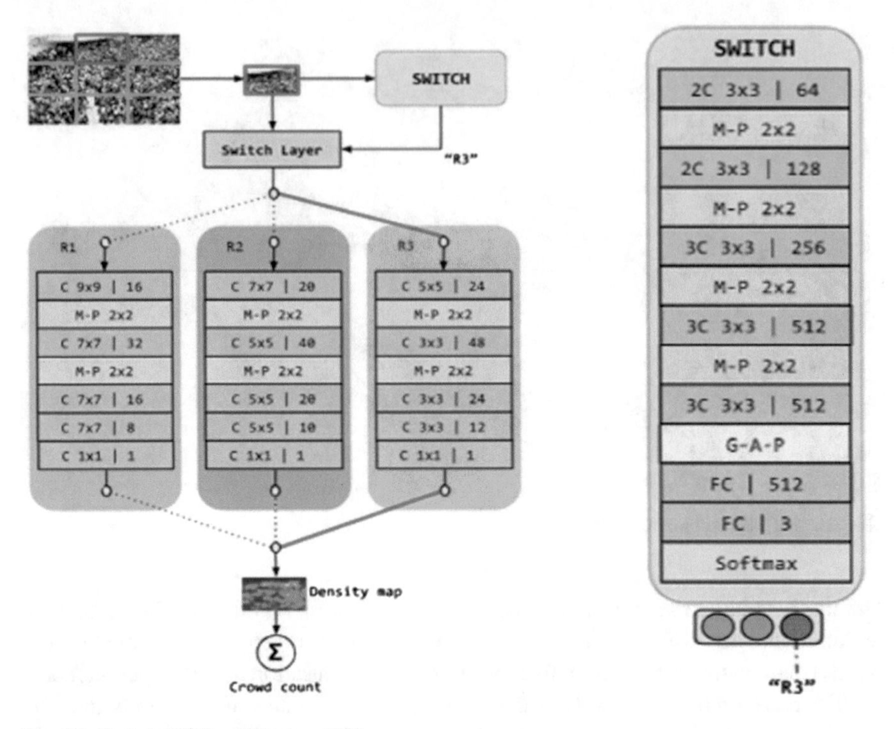

Fig. 17 Switch CNN architecture [16]

Each CNN regressor is trained to learn features that are adapted to a particular scale. Architecture of switch CNN is illustrated in Fig. 17. As we can see in the figure, a red color patch from the grid of crowd scene image is relayed to the R3 regressor by using a switch based upon the density level of the patch. The regressor predicts the corresponding density map of the patch that is summed upon to estimate crowd count. Each CNN regressor has four conv layers and two max pooling layers.

The switch consists of switch classifier and switch layer. Switch classifier determines the regressor to which patch is to be relayed and switch layer relays the patch to the correct regressor as inferred by switch classifier.

3.3 Cascaded CNN

Sindagi and Patel [17] proposed an approach for density estimation and crowd count classification (high level prior) together. The CNN network can learn discriminative features of crowd and to estimate count of people in the entire image irrespective of scale variations. The high-level prior stage for estimating count is jointly learned with density estimation using a cascade of CNN network as shown in Fig. 18. Both the

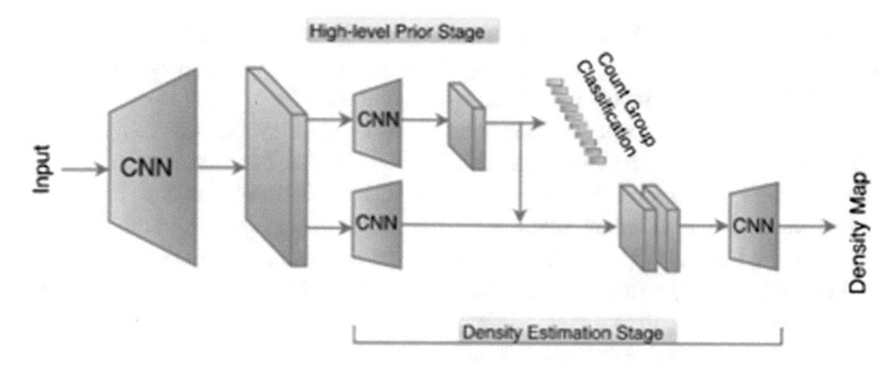

Fig. 18 Joint learning of high-level prior stage and density estimation stage [17]

task viz. counts classification and density map share some common layers of convolution network and followed by two parallel CNN networks specifically designed for individual tasks.

The high-resolution density map is obtained by concatenating global features learnt by high level prior and feature map obtained from the second set of layers. The concatenated features are processed by a set of stride convolutional layers to produce a high-quality density map. Cascaded CNN architecture is shown in Fig. 19. The architecture consists of three stages, shared convolution layers, high level prior stage and density estimation stage. The initial shared network consists of only two conv layers with parametric Relu(PReLU), first conv layer consists of 16 filters of size 9×9 and second conv layer consists of 32 filters of size 7×7. The second stage i.e. high-level prior stage consists of 4 conv layers, each layer followed by PReLU. Conv

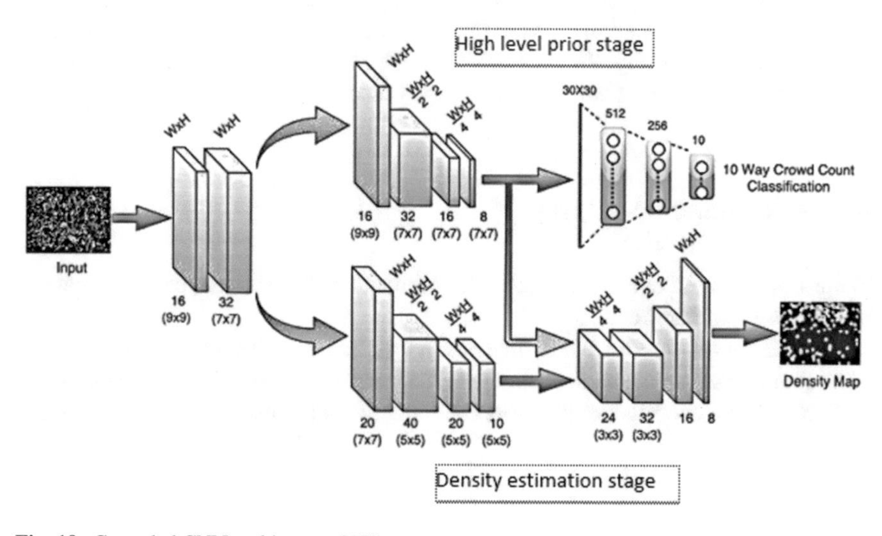

Fig. 19 Cascaded CNN architecture [17]

layers are followed by three FC layers having 512, 256 and 10 neurons respectively indicating count class of input image. Spatial pyramid pooling is added after the last conv layer to feed FC layers with fixed size output images.

The third stage i.e. density estimation consists of 4 conv layers followed by PReLU after every layer. The first conv layer has 20 filters of size 7×7, second conv layer has 40 filters of size 5×5, third conv layer has 20 filters of size 5×5 and last conv layers have 10 filters of size 5×5. The output of this stage is combined with output of the last conv layer of high-level prior stage by using a set of two conv layers and two fractionally stride conv layers. Fractionally stride conv layer is used to up sample the feature map to image original size to restore the details lost due to down sampling by max pool layers used initially.

3.4 CSRNet

Li et al. [18] proposed a deep learning network to estimate people's count and high-quality density map in highly congested regions. The approach consists of two major components: front end CNN for 2D feature extraction and back end CNN that uses dilated kernel for larger receptive fields to cope up with highly congested scenes and to replace pooling operation. VGG16 architecture is used in the frontend network because of its flexible architecture and its strong transfer learning capability with the backend network. Fully connected layers of VGG16 that are used for classification are removed to build proposed CSRNet. The first 10 layers of VGG16 and three max pooling layers are used in the front-end network as shown in Fig. 20. To cope up with down sampling issues linked with size of output and to maintain output resolution for high quality density maps, dilated conv layers are used in backend networks. Four different configurations having different dilation rate are used in the backend network as shown in Fig. 20 i.e. CSRNetA, CSRNetB, CSRNetC, CSRNetD with dilation rate of 1, 2, 4, combination of 2 and 4 respectively.

4 Performance Analysis

Table 1 provides an overview of these four models with their year, category, dataset used and salient features.

Table 2 provides performance analysis in terms of estimation error (MAE, MSE) of all four models on the most challenging dataset used for crowd counting purposes.

Table 3 is an overview of the most challenging dataset used for crowd counting.

Configurations of CSRNet			
A	B	C	D
input(unfixed-resolution color image)			
front-end (fine-tuned from VGG-16)			
conv3-64-1 conv3-64-1			
max-pooling			
conv3-128-1 conv3-128-1			
max-pooling			
conv3-256-1 conv3-256-1 conv3-256-1			
max-pooling			
conv3-512-1 conv3-512-1 conv3-512-1			
back-end (four different configurations)			
conv3-512-1	conv3-512-2	conv3-512-2	conv3-512-4
conv3-512-1	conv3-512-2	conv3-512-2	conv3-512-4
conv3-512-1	conv3-512-2	conv3-512-2	conv3-512-4
conv3-256-1	conv3-256-2	conv3-256-4	conv3-256-4
conv3-128-1	conv3-128-2	conv3-128-4	conv3-128-4
conv3-64-1	conv3-64-2	conv3-64-4	conv3-64-4
conv1-1-1			

Fig. 20 CSRNet network architecture [18]

5 Conclusion

In this chapter, we explored different ways of crowd counting from images and conclude that Convolution Neural Network models gives more accurate count as compared to traditional techniques of counting based upon hand crafted features. Further, suitability of CNN architecture to crowd counting problem is discussed. We have explored four major CNN architectures used for crowd counting viz. Multi-column CNN, Switch CNN, Cascaded CNN and CSRNet. Performance analysis of four models over five different types of datasets containing images of varying density (high, medium, and low) is also presented. On exploring different architectures of CNN, we concluded that deeper networks provide more accurate results as compared to traditional methods of crowd count. We also conclude that CNN results depend

Table 1 Overview of CNN models

S.No.	Paper reference	Year	Category	Datasets	Features
1	Zhang et al. [15]	2016	Multitask CNN	ShanghaiTech, UCF_CC_50, UCSD, MALL, World Expo'10	Multitask multicolumn CNN focusing on scale variation to obtain high quality density map
2	Sam et al. [16]	2017	Patch based CNN	ShanghaiTech, UCSD, UCF_CC_50, WorldExpo'10	Three regressors having different filter sizes are trained on low, medium and high density image patches
3	Sindagi and Patel [17]	2017	Whole image-based CNN	ShanghaiTech, UCF_CC_50	Multitask cascaded CNN to learn crowd density and crowd classification
4	Li et al. [18]	2018	Context based CNN	UCF_CC_50, STA, STB, WorldExpo'10	Dilated convolution network with multiscale contextual information

upon resolution of image, physical conditions such as illumination level, camera angle and occlusion level. However, results of CSRNet are most promising out of the compared CNN architectures.

Table 2 Performance analysis

		Estimation error							
S.No.	Dataset	MAE				MSE			
		Multicolumn CNN	Switching CNN	Cacaded CNN	CSRNet	Multicolumn CNN	Switching CNN	Cacaded CNN	CSRNet
1	ShangaiTechA	110.2	90.4	101.3	**68.2**	173.2	135	152.4	**115**
2	ShangaiTechB	26.4	21.6	20	**10.6**	41.3	33.4	31.1	**16**
3	UCF_CC_50	377.6	318.1	322.8	**266.1**	509.1	439.2	397.9	**397.5**
4	WorldExpo'10	11.6	9.4	8.86	**8.6**	–	–	–	–
5	UCSD	**1.07**	1.62	1.51	1.16	**1.35**	2.1	–	1.47

Bold numerals signifies the minimum MSE and MAE on various datsets compared for different models

Table 3 Dataset overview

Dataset	Image resolution	No. of images	Maximum crowd count	Minimum crowd count	Average crowd count	No. of labelled people	Remarks
ShangaiTechA	Variable	482	3139	33	501.4	241,677	Images selected from internet, more dense images
ShangaiTechB	768 × 1024	716	578	9	123.6	84,488	Images collected from Shangai, low density images having varying scale and perspective
UCSD	158 × 238	2000	46	11	24.9	49,885	Low density images collected from video camera
UCF_CC_50	Variable	50	4543	94	1279.5	63,974	High range of density images from diverse scenarios
WorldExpo'10	576 × 720	3980	253	1	50.2	199,923	Low density images specific for cross scene counting

References

1. Wang H, Cruz-Roa A, Basavanhally A, Gilmore H, Shih N, Feldman M, Tomaszewski J, Gonzalez F, Madabhushi A (2014) Cascaded ensemble of convolutional neural networks and handcrafted features for mitosis detection. In: Medical imaging 2014: digital pathology, vol 9041, p 90410B
2. Wang H, Cruz-Roa A, Basavanhally A, Gilmore H, Shih N, Feldman M, Tomaszewski J, Gonzalez F, Madabhushi A (2014) Mitosis detection in breast cancer pathology images by combining handcrafted and convolutional neural network features. J Med Imaging (Bellingham) 1(3):034003

3. Dollár P, Wojek C, Schiele B, Perona P (2012) Pedestrian detection: an evaluation of the state of the art. IEEE Trans Pattern Anal Mach Intell 34(4):743–761
4. Chan AB, Vasconcelos N (2012) Counting people with low-level features and Bayesian regression. IEEE Trans Image Process 21(4):2160–2177
5. Chen K, Loy CC, Gong S, Xiang T (2012) Feature mining for localised crowd counting. In: Proceedings of the British machine vision conference 2012, pp 21.1–21.11
6. Giuffrida MV, Minervini M, Tsaftaris S (2015) Learning to count leaves in rosette plants. In: Proceedings of the computer vision problems in plant phenotyping workshop 2015, pp 1.1–1.13
7. Wu X, Liang G, Lee KK, Xu Y (2006) Texture analysis and learning*
8. Lempitsky V, Zisserman A (2010) Learning to count objects in images. In: Advances in neural information processing systems 23: 24th annual conference on neural information processing systems 2010, NIPS 2010, pp 1–9
9. Rabaud V, Belongie S (2006) Counting crowded moving objects Cited by me. In: Computer vision and pattern recognition, 2006. IEEE
10. Schölkopf B, Platt J, Hofmann T (eds) (2007) Fast discriminative visual codebooks using randomized clustering forests. In: Advances in neural information processing systems 19: proceedings of the 2006 conference. The MIT Press
11. Duygulu P, Barnard K, de Freitas JFG, Forsyth DA (2002) Object recognition as machine translation: learning a lexicon for a fixed image vocabulary. In: Heyden A, Sparr G, Nielsen M, Johansen P (eds) Computer vision—ECCV 2002: 7th european conference on computer vision Copenhagen, Denmark, may 28–31, 2002 proceedings, part IV, vol 2353. Springer, Heidelberg, pp 97–112
12. Surasak T, Takahiro I, Cheng C, Wang C, Sheng P (2018) Histogram of oriented gradients for human detection in video. In: 2018 5th International conference on business and industrial research (ICBIR), pp 172–176
13. Lowe DG (2004) Distinctive image features from scale-invariant keypoints. Int J Comput Vis 60(2):91–110
14. Wu B, Nevatia R (2005) Detection of multiple, partially occluded humans in a single image by Bayesian combination of edgelet part detectors. In: Tenth IEEE international conference on computer vision (ICCV'05), vol 1, pp 90–97
15. Zhang Y, Zhou D, Chen S, Gao S, Ma Y (2016) Single-image crowd counting via multi-column convolutional neural network. In: 2016 IEEE conference on computer vision and pattern recognition (CVPR), pp 589–597
16. Sam DB, Surya S, Babu RV (2017) Switching convolutional neural network for crowd counting. In: IEEE conference on computer vision and pattern recognition (CVPR), pp 4031–4039
17. Sindagi VA, Patel VM (2017) CNN-based cascaded multi-task learning of high-level prior and density estimation for crowd counting. In: 2017 14th IEEE international conference on advanced video and signal based surveillance (AVSS), pp 1–6
18. Li Y, Zhang X, Chen D (2018) Csrnet: dilated convolutional neural networks for understanding the highly congested scenes. In: IEEE/CVF conference on computer vision and pattern recognition, pp 1091–1100
19. Zhang C, Li H, Wang X, Yang X (2015) Cross-scene crowd counting via deep convolutional neural networks. In: IEEE conference on computer vision and pattern recognition (CVPR), pp 833–841

AI and IoT for Intelligent Transportation

IoT Based Regional Speed Restriction Using Smart Sign Boards

P. Madhumathy, H. K. Nitish Kumar, Pankhuri, and D. S. Supreeth Narayan

Abstract Major cause for fatal accidents on the road is over speeding. Accident risk increases with an increase in speed. The judging ability of upcoming events also gets declined while moving at higher pace, which causes judgment mistakes and leads to a crash. Around 30% of road accidents are due to over speeding. There have been various ways to avoid accidents due to over speeding, but none of them can automatically control the speed and customize the regional speed limit together. An IoT-based smart solution is discussed to overcome this, limiting the vehicle's top speed to a particular region even though people are unwilling to use control stations, smart signboards, and speed control unit in the vehicle.

Keywords Fatal accidents · Over speeding, IoT · Control stations · Smart sign boards · Speed control unit

1 Introduction

Rash driving is a major moving traffic violation. It is driving without thinking about the safe driving. In the present scenario, rash driving and over speeding are the major traffic violations. When we investigated the road accident records, we determined that India heads the road accident list amid south-east Asian countries. Over speeding, rash driving, breaking of traffic rules, misconceive the sign boards, and sleepiness

P. Madhumathy (✉) · H. K. Nitish Kumar · Pankhuri · D. S. Supreeth Narayan
Department of Electronics and Communication Engineering, Dayananda Sagar Academy of Technology and Management, Bengaluru, India
e-mail: sakthi999@gmail.com

H. K. Nitish Kumar
e-mail: nitishardra@gmail.com

Pankhuri
e-mail: bhakti.pankhuri@gmail.com

D. S. Supreeth Narayan
e-mail: supreethnarayan@outlook.com

K. R. Ahmed et al. (eds.), *Deep Learning and Big Data for Intelligent Transportation*, Studies in Computational Intelligence 945,
https://doi.org/10.1007/978-3-030-65661-4_10

are main elements about the road accidents. We observe that when travelling during night time, the driver may skip the traffic signals. Failure to identify traffic signals may be calamitous to the chauffer along with other passengers. A number of tough laws have been introduced however it could not control hard driving [1].

Till now, various ways have been developed to avoid accidents due to over speeding like sign posts which indicate the particular speed limit, attaching speed limiters to automobile engines that limit the opening of the inlet valve, GPS based speed indication, and RFID based speed control. But the major drawbacks in these methods are that it can either limit the speed to a particular value, or just indicate the speed limit, but the region's speed limit cannot be varied based on factors like weather and road condition [2].

A better was to overcome all these issues is to take control of the vehicle when the riders do not abide by the rules. In this way, there is no need to impose stringent laws and more importantly, it will drastically reduce the accidents or at least the severity of the accidents [3].

Through this project, we mainly intend to create a safe road network by proposing a sagacious method to avoid accidents due to over speeding. The need for the hour is to develop a cost-effective system that can be installed in vehicles. For the time being, we have developed a system for electric vehicles, but a similar system with slight modifications can be implemented for gasoline powered vehicles.

As the main goal to prevent the occurrence of accidents due to over speeding and overtaking of vehicles, the following ideas were found which were found similar and useful for our project. This paper provided an idea of having a transmitting device within a signpost, indicating the pre limit to the chauffeur and a receiver device equipped within the vehicle [4, 5]. Another idea is the emergency button or a switch provided in the vehicle for emergency conditions. But the limitation is the usage of RFID devices which has less accuracy for fast-moving vehicles [6]. This paper provides the usage of BLE devices for information broadcast to vehicles. Though it could detect any type of signboard, it would only indicate the speed limit to the driver [7]. The idea of RF transmitter and receiver is the same as seen before but the idea of GSM usage to alert the police about the violation. But the interference by other vehicles reduces the efficiency of the RF system [8]. The concept of Road Side Unit (RSU) and On-Board Unit (OBU) has been taken from this paper where the RSU is connected to a remote server. The OBU collects various data about the vehicle like current speed, vehicle ID, etc. and transmits it to the RSU. This system only just used for monitoring traffic violations [9].

2 Background

In 2015, considering road safety as prime concern, India approved the Brasilia declaration and pledged to reduce road fatalities by 50%. Till now, the reduction in road accidents is not much considerable. During the last 10 years, the par death count stands at 1.5 lakh for an annum. And if this scenario continues, by 2020 road fatalities would become third major contributor to global cause of injury and disease. Increase in par speed of the vehicle is proportional to the rise of occurrence of an accident and also the austerity of the corollary of the accident. For an instance, 1% increase of mean speed contributes 4% increase in accident risk [10, 11].

India has a large count of unsafe roads with one death case in every five minutes. And this could rise by 2020 without any changes in present driving methods. Majorly, people of age around 15–44 are deceased in 50% of road accidents which occur worldwide. Road accidents are foreseeable and avertible, by enforcing suitable safety interventions. With the ever-expanding population, the dependency of vehicles leads to an increase in traffic and therefore increases in road fatalities [12]. The prevailing requirement is to establish a cost-effective structure that can be set up in any kind of vehicles to help prevent a crash or at least reduce the effects of accidents thus, improving road safety.

3 Problem Statement

"Speed thrills, but kills". This slogan has been used in many road safety placards. Increasing road traffic accidents is a matter of great concern today. Under such circumstances, improving road safety by adopting suitable safety measures is necessary. Statistics show that over 64.4% deaths are caused due to road accidents, mainly over speeding and wrongful over taking. Many strict laws are enforced but harsh driving is still not under control [13].

Various ways have been developed to avoid accidents due to over speeding which includes:

(1) Sign posts which indicates the particular speed restriction in that range or highway.
(2) Speed limiters attached to the automobile that limits the opening of the engine's inlet valve.
(3) GPS based speed indication.
(4) RFID based automatic speed indication.

But the major drawbacks in the above methods are that it can either limit the speed only to a particular value or just indicate the driver about the speed limit, but none of them control the speed depending on the regional conditions [14–17].

3 Objectives

- The need for the hour is to develop a cost-effective system that can be installed in any kind of vehicle to avert a crash its effects, thus improving road safety.
- Avoid accidents due to overtaking.
- Vary the speed limit of a region depending on the conditions like weather, time, and traffic density.

4 Literature Survey

Vengadesh and Sekhar [8]. Here the aim was to automatic speed control of vehicle in restricted areas such as schools, parks, and hospitals where it is necessary to control the automobile speed. The chauffeur has no control in such situations; automatic controls are grasped by utilizing electronic systems which makes the driver forcefully abide by the speed limit. This project uses an RF transmitter for speed limit indication. It is installed at the front and rear end of the restricted areas. An RF receiver is mounted within the vehicle. Odometer acquires the vehicle speed and transfers it to the controller. The present speed of vehicle is tracked by ultrasonic sensor and is also delivered to the controller. The controller compares both acquired values and if the driver tires to enhance the speed, controls are taken by the electronic controller. If the driver forces to enhance the speed then driver details are forwarded to the nearest traffic police station. This detail contains the violated speed and vehicle registration nuber. The data is transmitted with the help of GSM module [18, 19].

Satyanarayana et al. [6]. The RF transmitters are installed at the entry and exit of the zones and RFID receiver are been placed inside the vehicle. The odometer records the vehicle speed and the controller compares current speed with the speed limit. If speed of the vehicle goes beyond limit then speed is reduced automatically. In case of emergency, then a switch is accessible in the automobile. During the switch is turned ON, the speed is not controlled automatically. When the person turns the switch ON, then RF receiver doesn't work and the speed is controlled by driver. The cloud (Blynk) stores the driver and vehicles details which are further used to keep a track on the vehicle route by Blynk application. Using this application distinct token number is generated for each vehicle.

Volam et al. [7]. Bluetooth Low Energy (BLE) is used to send traffic sign board data to automobiles. These various signs are categorised, corresponding to the local traffic laws to assist the chauffeur. An algorithm is used to locate the required sign board nearly based on a data acquired by four Bluetooth radios mounted inside the vehicle. In addition, it is also equipped with a multi-level alerting setup to categorise the signs determined by the automobile.

Anushya [9]. Conveyance tracking as part of V2I assists reduction in the difficulties led by vehicles, like traffic jam, violation of rules, and road fatalities. To regulate these, V2I communication is used between vehicle and traffic light pole. A prototype of On-Board Units (OBU) and Road Side Units (RSU) are created. Vehicle uses Zig-Bee technology to transmit details like current speed, position and registration number [20]. Data related to the user's driving is sent to a tracking server by the RSU to charge penalty on offender. CAN bus is used for communication between the modules present in the vehicle. OBU collects the information using this bus and transfer the details to the RSU in real time. The server after analysis sends the notification to the driver using E-mail service.

5 Proposed Methodology

The system outline is shown in Fig. 1. The system is made of 3 major units.

1. The Road Side Unit.
2. The Speed Control Unit and the Video Streaming Unit. The SCU and VSU together are known as the On Board Unit (OBU).
3. Monitoring Server and Database.

The monitoring server and database are used to set regional speed limit. The database is a collection of API keys and location information. Every time the monitoring server receives a request from a sign board, the API key of the sign board is mapped to the sign board location. After this, the server returns the speed limit and distance till which the vehicle has this speed limit (Fig. 2).

The road side unit is nothing but a smart sign board. It is utilized to acquire information from the monitoring server and broadcast it to the vehicles within its vicinity.

The speed control unit consists of a receiver, which receives values transmitted by the sign board. By comparing the received values and current values, the controller in this unit takes decisions for controlling the motor driver in order to maintain the speed within the limit.

The video streaming unit consists of an obstacle detector, camera, and a display. Here, obstacle means a vehicle at the rear end trying to overtake. This unit assists us in overtaking large vehicles which are travelling at a low speed.

Parts of the System

- Sensors

1. Speed Sensor
2. Ultrasonic Sensor

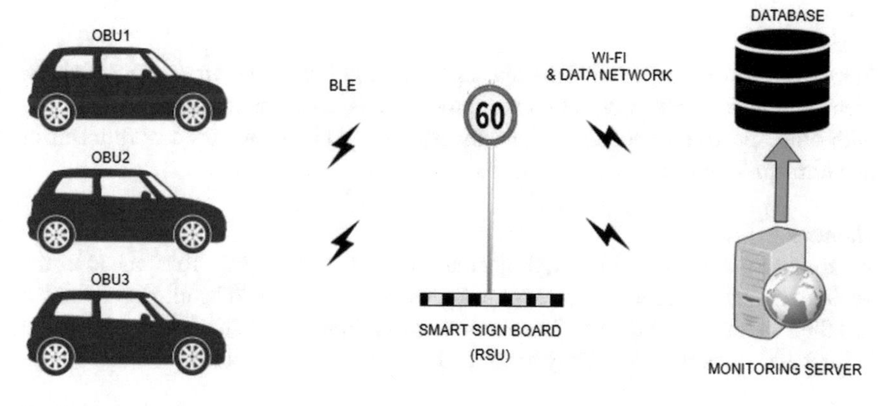

Fig. 1 Design of the system

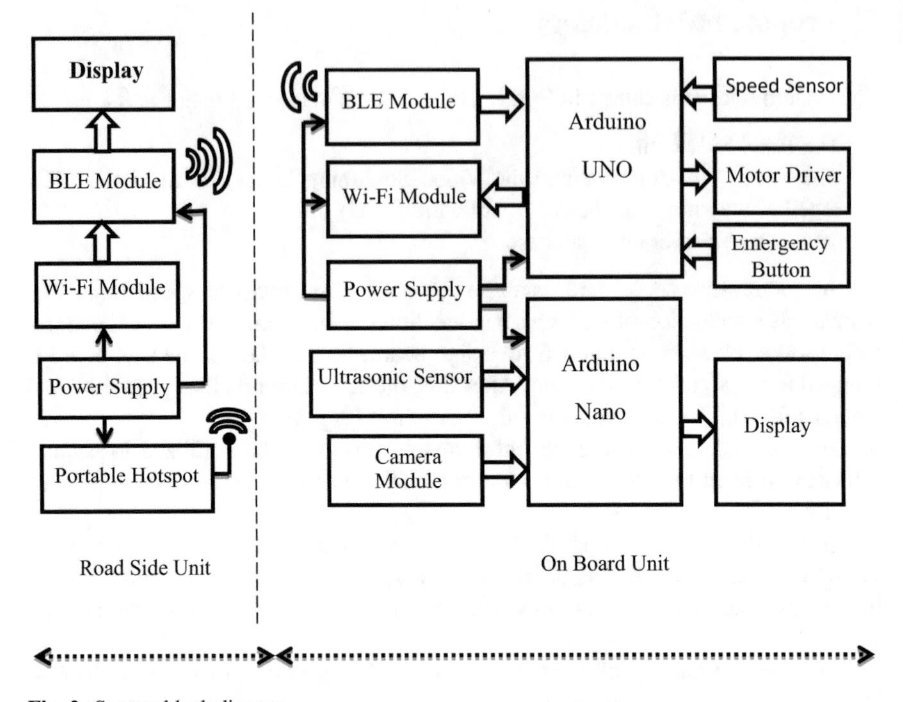

Fig. 2 System block diagram

- Camera
- Display Module
- Arduino UNO
- Arduino Nano
- BLE Module
- Wi-Fi Module & Portable Wi-Fi Hotspot.

Sensors

Sensors are devices which find certain action or variations in its surrounding environment. It transfers this information to a processor. A sensor is frequently used along with other electronic modules. The sensors equipped here are speed or rpm sensor and ultrasonic sensor.

Camera

A camera module is used in the project as a part of recording the road ahead of the vehicle. It starts recording once a vehicle presence is confirmed by ultrasonic sensor installed behind the vehicle. The recording continues until there will be vehicle behind and stops recording only when there are no vehicles behind.

Display Module

The display module used in the project as a part of displaying the streaming video captured by the camera. This module helps the vehicles behind to view road beyond the current vehicle for safe overtake.

Arduino Uno

The Arduino Uno is a microcontroller board established on the ATmega328. It has fourteen digital input/output pins, 6 analog input pins, a 16 MHz ceramic resonator, a USB connection, a power jack, an ICSP header, and a reset button. It contains all things needed to backing the microcontroller; just interface it to a computer with the help of USB cable or power it with a AC-to-DC adapter or battery to start. Arduino UNO is used to control motor driver, read data from speed sensor, communicates in serial way with Bluetooth module.

Arduino Nano

The Arduino Nano is a small, breadboard-friendly board based on the ATmega328 (Arduino Nano 3.x), it has the same functionality of the Arduino Duemilanove, but in a different package. It lacks only a DC power jack, and works with a Mini-B USB cable instead of a standard one. This microcontroller is used to control camera and display module.

BLE Module

A BLE (Bluetooth low energy) is a transceiver that is used for communication between two or more devices simultaneously. A transceiver which means both sending and receiving of data is used here for infrastructure to vehicle communication, and the transceiver operations are performed and controlled by the microcontroller based instructions.

Wi-Fi Module

The Wi-Fi module is a microchip, with full TCP/IP stack and microcontroller capability. The usage of this module in this project is to act as both client and server at two different areas. At the RSU end, the Wi-Fi module acts as client and requests data from the cloud server which is later sent to the BLE module. At the OBU end, the Wi-Fi module acts as the server and shares the data with mobile.

Unit Flow

The functionalities of each of the units are shown below in terms of flow charts.

The flow of control in RSU is shown in Fig. 3. As soon as this unit is powered up, it sends the (Application Programming Interface) API key to the server. The control station receives the API key. Each API key corresponds to a particular region. Then, the server returns the speed and distance limits back to the sign board. Now, the connection is established and the speed limit is displayed. If the sign board doesn't receive the values within a specified time, the sign board will send its API key once again. The sign board now advertises its Bluetooth Low Energy (BLE) ID within

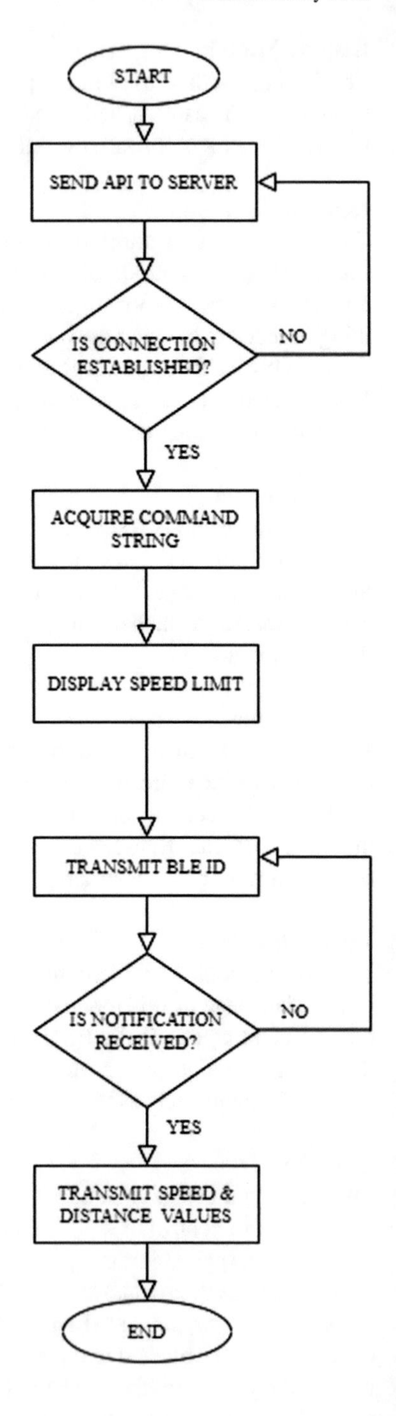

Fig. 3 Algorithmic flow chart of the Road Side Unit

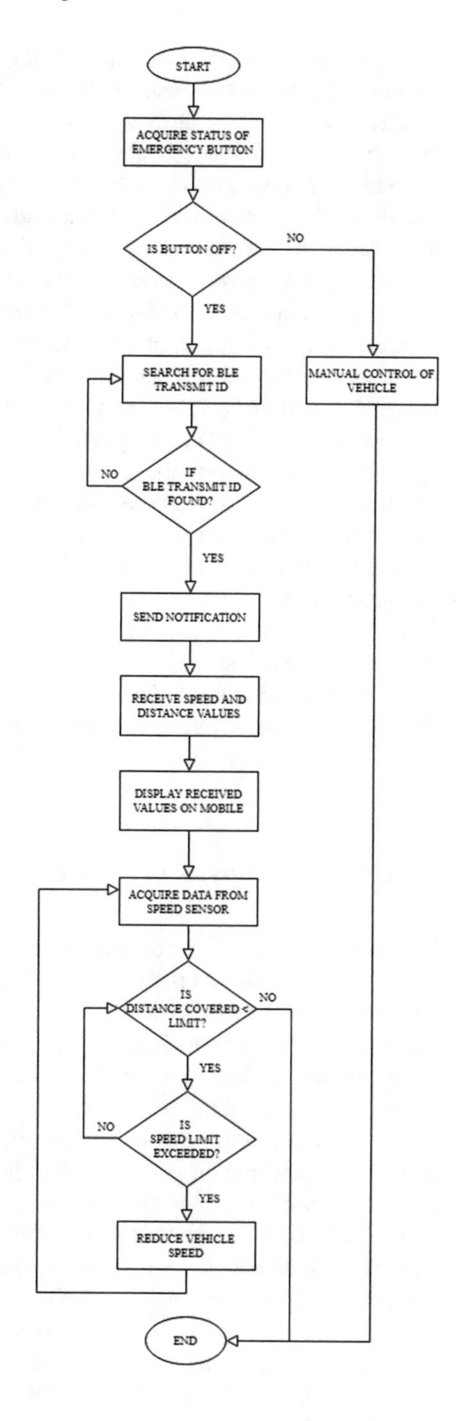

Fig. 4 Algorithmic flow chart of the Speed Control Unit

its range and waits for a notification from any vehicle. If no, it re-advertises. If yes, transmits the speed and distance limits to the vehicle.

The flow of control in Speed Control Unit (SCU) is shown in Fig. 4. First, the emergency button status is checked. If it is on, there are no speed and distance restrictions imposed on the vehicle. This happens in case of an emergency. If the button is off, it searches for a sign board advertisement. Once it passes by a smart sign board along the road network, it notifies the sign board and connection is established. The SCU in the conveyance accepts the speed and distance limits from RSU and the speed limit value can also be viewed in the smart phone. The controller in the SCU compares this received limit with the current speed of the vehicle acquired by the speed sensor. If the current speed is less than or equal to the restricted speed, no action is taken. If current speed is greater than the restricted speed, then the speed of the conveyance is reduced in a loop, until it is equal to or below the limiting value. This speed is maintained until the distance limit is exceeded after which the vehicle does not have any speed or distance restriction. While the vehicle is within the distance limit, the driver can travel at a lower speed also.

The Video Streaming Unit (VSU) is designed to prevent accidents due to overtaking and the flow is shown in Fig. 5. The ultrasonic sensor continuously checks for an obstacle (vehicle within limit), if found, the controller acquires the video from the camera at the front end of the vehicle and displays it using the LCD screen at the rear end. The video is streamed for ten seconds. The control is transferred back to the ultrasonic sensor for vehicle detection. If there is no vehicle behind, the camera and display will be in sleep mode (Figs. 6, 7 and 8).

6 Circuit Diagram of the Proposed System

The system in this project contains two major units. First is the Road Side Unit (RSU) and the second is the On Board Unit (OBU). Once the Road Side Unit is powered up, it connects to the internet with the help of Wi-Fi module and acquires speed limit and distance values from data base. Once the value is received, the RSU's BLE module starts to transmit data continuously. When the vehicle with OBU installed start, the microcontrollers and the BLE module are powered up.

With the help of a common voltage source supply and a common ground all the devices are powered up from the controller. Each sensor that is connected collects data in either of the two forms, analog value or digital value.

OV7670 camera module is a small size, low voltage, and highly sensitive CMOS image sensor module to capture and process the image. Using the SCCB bus control, the sensor outputs thee frame, sampling, and many resolutions of 8 bits of data. The VGA can reach up to 30 FPS. We can control the image quality, format, and mode of transmission. Supports image formats like VGA and CIF. Also provides saturation level auto adjust and edge enhancement level auto adjust.

ILI9225B module displays video which is streamed from the camera module. This module is mounted on the back (rear end) of the vehicle, facilitating the other vehicle

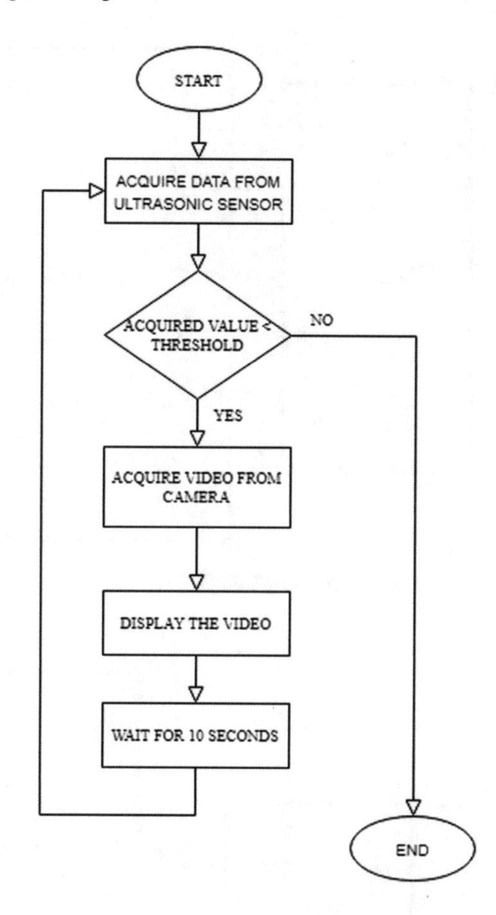

Fig. 5 Algorithmic flow chart of the Video Streaming Unit

behind to view the road ahead. This helps the drivers to decide when to overtake. It is powered up only when other vehicle presence is confirmed by controller. The TFT liquid crystal display has a resolution of 176 * 220 dots.

HC-SR04 module is an ultrasonic ranging module which provides a non-contact measuring range from 2 to 400 cm, with an accuracy of up to 3 mm. The module consists of a ultrasonic transmitter, a receiver and a control circuit. The module is installed at the rear end of the vehicle. Once the system is powered up, the sensor checks for the presence of any vehicle near to it. When vehicle presence is detected, that is when the measured value of sensor is less than the threshold value; the microcontroller turns the camera and TFT display modules.

TM1637 is a 4 bit digital LED display module. It is an affordable solution to display the output data. The 7 segment LED display has 4 digits. It has a I2C bus which can be controlled using only 2 wires. The hardware connection includes two signal connections CLK and DIO.

LM393 speed sensor is used for detecting the speed of the motor. It consists of IR transmitter and receiver, and a comparator. The basic operation of the sensor

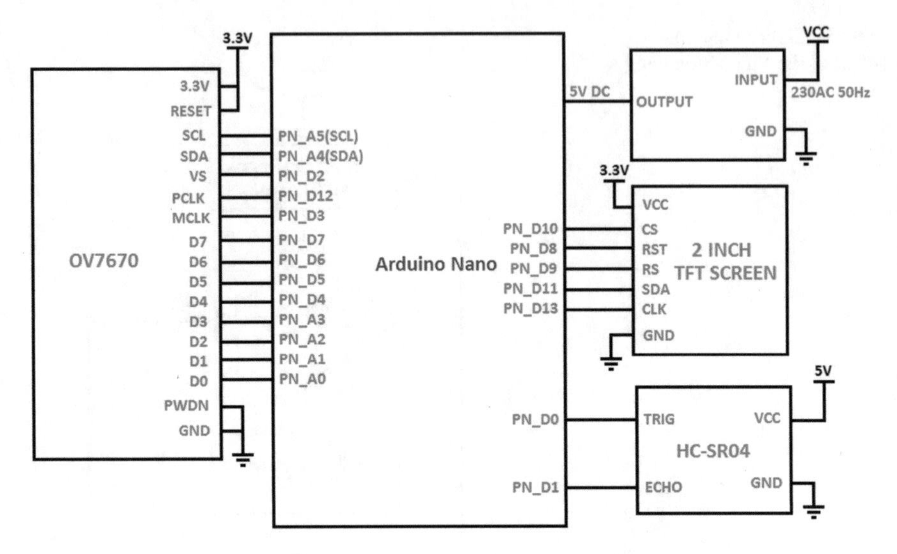

Fig. 6 Video Streaming Unit (VSU)

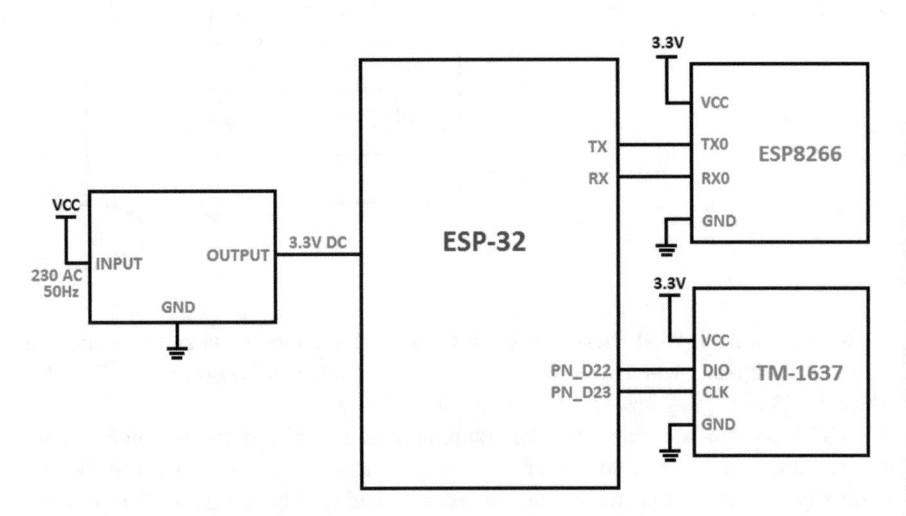

Fig. 7 Road Side Unit (RSU)

is—when anything passes between the sensor slots, it generates a digital pulse on the pin D0. The pulse is a digital TTL signal. This pulse is converted to RPM by the micro-controller. The distance travelled by the car is also calculated using these pulses.

ESP32 consists of a 2.4 GHz Wi-Fi, Bluetooth and BLE on a single chip. In this project, ESP32 module is mainly used for communication between RSU and OBU. This communication involves transfer of speed and distance limits. The RSU ESP32 module receives data from the Wi-Fi module through UART interface, and

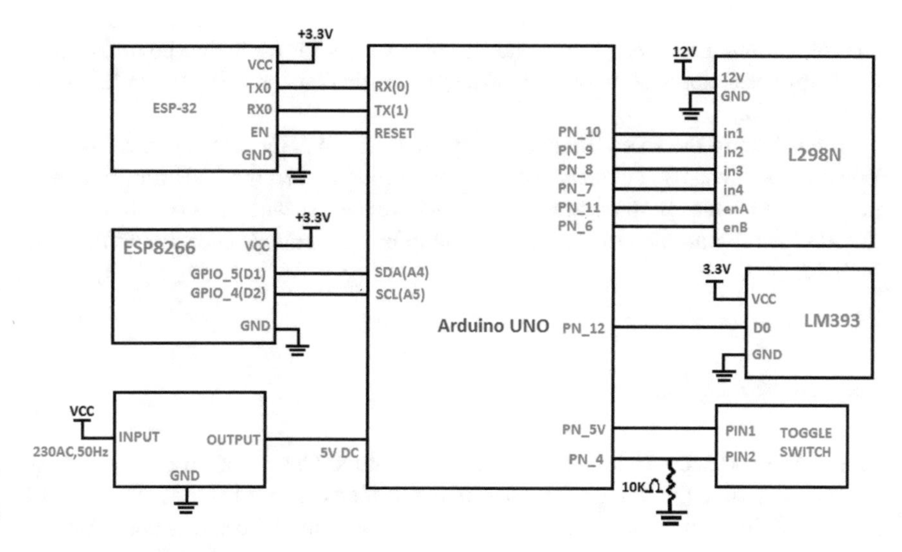

Fig. 8 On Board Unit (OBU)

then transmits them in the range. The OBU ESP32 module receives this data and forwards to the controller through UART interface.

ESP8266 is a 2.4 GHz Wi-Fi module. In this project, this module is used to get data from the database of the control station to the OBU. Another Wi-Fi module used in the vehicle unit is used to host a web-server which displays the speed limit to the rider.

L298N module is a motor driver. It is used to control the speed and direction of the motors using Pulse-Width Modulated signals. Here is used to reduce the speed gradually using varying duty cycles of the PWM waves given to the DC BO-motors. The speed reduction takes place only if the speed of the vehicle is greater than speed limit indicated and provided by the sign board.

7 System Implementation

7.1 System Prototype

The system implementation is done using a hardware prototype. Arduino's IDE is used for programming all the controllers used in this project. ThingSpeak IoT platform is used for creating a monitoring server.

A control station covers a large geographical area and it consists of a monitoring server. All the sign boards in this geographical area are controlled by the control station. The speed limits and distance values for each sign board is set in the ThingSpeak application by monitoring the traffic density, weather condition and road condition.

The values in the road side unit are updated periodically by the control station. When the RSU has lost contact with the monitoring server, the RSU transmits pre-programmed values till the connection is back. In regions with no access to Internet, the RSU is programmed to change the data to be transmitted according to the time in that region.

7.2 Motor Driver

Motor driver is for controlling the speed of the motors. The motors in electric vehicle are powered using DC voltage. In order for controlling the speed of the motors, PWM signals will be used. Figure 9 shows relation between duty cycle and average voltage. The controller in the speed control unit maps the speed range of the vehicle with the PWM range depending on its resolution. Later, the current speed and the speed limit are converted to PWM values. The controller compares the two speeds in terms of these PWM values and accordingly gives control signals to the motor driver. After comparing, if current speed is higher than the speed limit, motor driver lowers the average DC voltage by reducing the width of the pulse (duty cycle). The pulse width of supply voltage to the motor is reduced continuous in a loop till the present speed is <= speed limit. Here, even though the driver tries to raise the accelerator, the controller and the motor driver together won't allow the speed of the vehicle to increase.

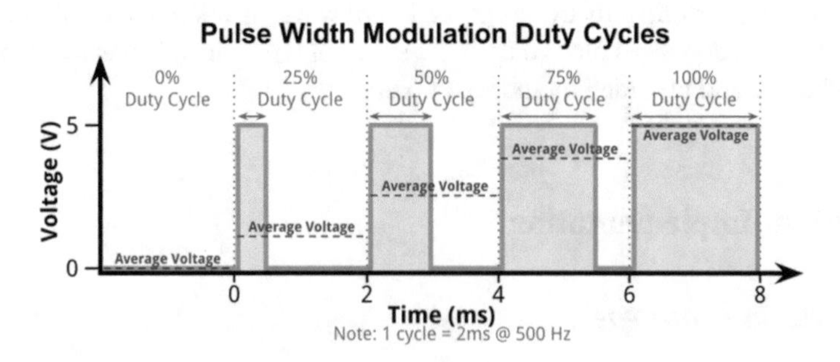

Fig. 9 Relation between pulse width and average voltage

7.3 Bluetooth Low Energy

BLE modules are used in RSU and OBU to transmit and receive data. The BLE in the RSU is known as Server and the BLE in a vehicle is known as client. The reasons for choosing BLE over RFID are as follows:

1. Low power consumption
2. Better range
3. Less pairing time
4. Broadcast mode is possible.

For establishing a connection between a client and a BLE server, the server continuously advertises the ID. Similarly, the client BLE will always be searching for the server ID. When the vehicle is in the range of the sign board, the client finds this ID and notifies the server BLE. The server BLE considers this as an acknowledgement and transmits the data to the vehicle.

7.4 BLE Packet Format

Figure 10 shows the packet formats for BLE advertisement and data transmission. The Preamble is used for synchronization between a client and a server. The access address is used for identifying a connection. The Protocol Data Unit (PDU) can be an advertising channel PDU or data channel PDU. CRC is used to detect the

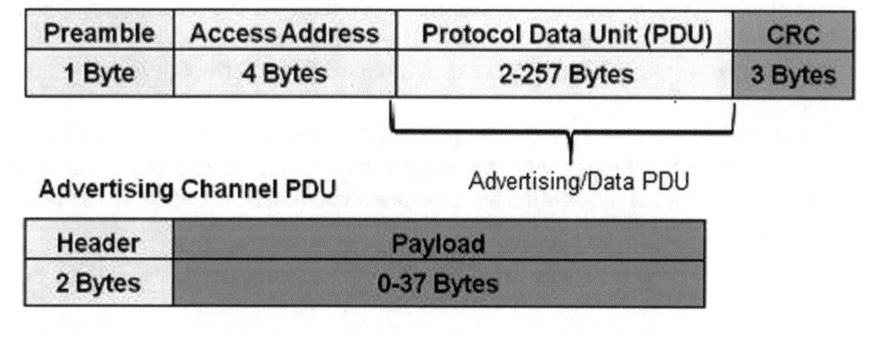

Fig. 10 BLE packet format

communication errors. The header in the PDU represents the type of connection that can be established between the devices. MIC stands for Message Integrity Check and it is optional.

8 Results and Discussion

8.1 Real Time Implementation

The project is implemented by constructing a prototype of a vehicle on a miniature 4 wheeler vehicle. All the sensors, controllers and other components are placed on a platform upon the prototype providing sufficient spacing for each without each other interfering in the working of the other. Here, the vehicle is powered by a rechargeable battery pack.

Similarly, a miniature smart sign board is constructed which consists of a controller, battery pack, Portable Wi-Fi hotspot, and a seven segment display to indicate the speed limit.

The working environment is set up in such a way that, first the vehicle travels for some distance at some random speed without encountering a sign board. After traveling 'x' meters, the vehicle passes by a signboard and picks up the data transmitted by the sign board. Depending on the initial speed, the vehicle may experience a deceleration if it is travelling above the speed limit. If the speed reduces below a limit, the Speed Control Unit does not allow itself to increase the speed until it crosses the distance limit. The controller in the vehicle is programmed in such a way that it performs all the above mentioned operations automatically when powered on.

Alternatively, the vehicle may be controlled wirelessly using a Bluetooth, which would be more realistic and analogous to a rider riding the vehicle.

Also, a web-server is created is order to monitor the performance of the project via a smart phone.

Moving on to the smart overtaking part, an object is introduced in the vicinity of the ultrasonic sensor at the rear end of the vehicle so that the display turns on to stream the video signal transmitted by the camera via a controller at the front end of the vehicle.

8.2 Discussion

Road Side Unit
See Table 1.

Speed Control Unit
See Table 2.

Table 1 Working of RSU under various conditions

Condition	Action
Notification from the vehicle	Transmit data (speed and distance)
Data from the control station	Update the display and data to be transmitted
Loss of connectivity to control station	Transmit pre-programmed values

Table 2 Working of SCU under various conditions

Condition	Action
Current speed greater than the speed limit	Speed is automatically reduced below the limit even though the rider tries to increase the speed and this speed is maintained till the distance limit is reached
Current speed less than or equal to the speed limit	No actions are taken. The SCU actively monitors the speed of the vehicle to make sure that the rider doesn't surpass the limit until the distance limit is reached
Emergency button turned on	Speed Control Unit will be turned off

Table 3 Working of VSU under various conditions

Condition	Action
Vehicle at the rear end, within the range of the ultrasonic sensor	Video streaming is on
Vehicle at the rear end, outside the range of the ultrasonic sensor	Video streaming is off

Video Streaming Unit
See Table 3.

Figure 11 shows how to set the speed and distance values in the form of a command string for particular sign board identified by its unique TalkBack ID. Upon clicking on the "Save" button next to the command string, the command string will be transferred to the respective sign board.

Figure 12 indicates the speed and distance values received by the vehicle from the sign board and also shows how the speed is reduced when present speed has crossed the limit. Once current speed falls down to the speed limit, the SCU doesn't allow the speed to exceed the limit. As soon as the distance limit is reached, the restrictions are lifted.

Figure 13 shows the hardware implementation of the Road Side Unit. It consists of a SSD which indicates the speed limit received from the control station and transmits this value to the vehicles on the road.

Figure 14 shows the working of VSU of the vehicle. The VSU is turned on when there's a vehicle at the rear end, within the range of the ultrasonic sensor.

Figure 15 shows the VSU in turned off condition when there's no presence of vehicle at the rear end. During this condition the TFT screen turns blank (white).

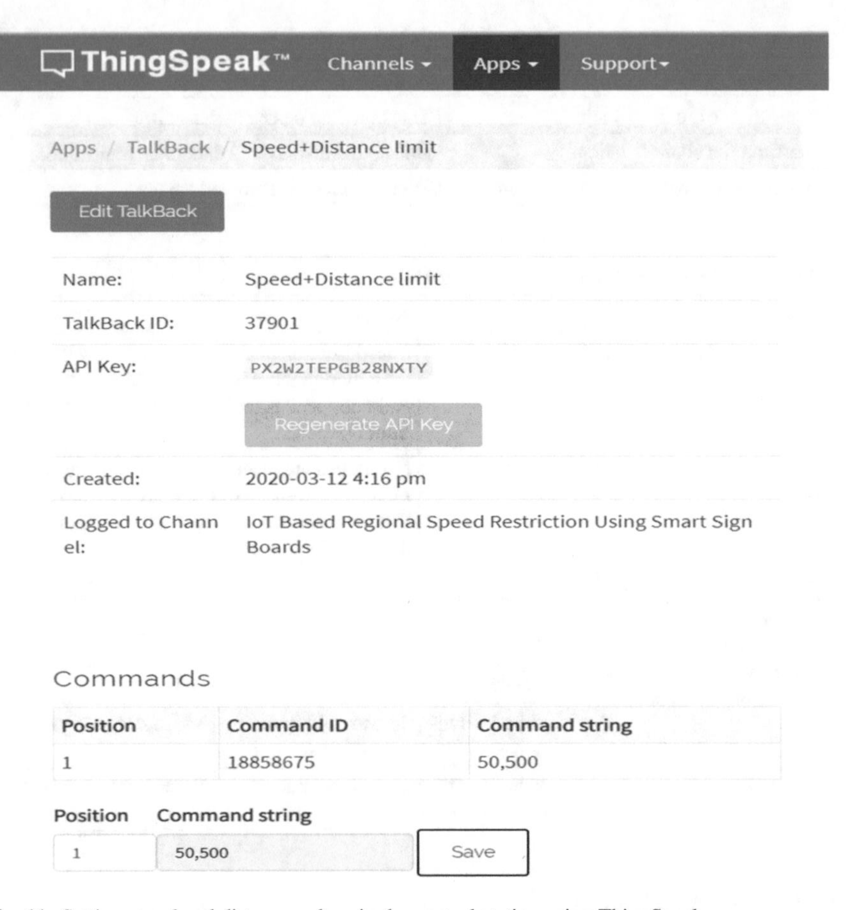

Fig. 11 Setting speed and distance values in the control station using ThingSpeak

Figure 16 shows the speed limit indicated to the driver through his smartphone connected to the web server hosted by the SCU.

9 Conclusion

Road accidents in India are one of the major sources of injuries, deaths and collateral damage each year. In 2018, a person died every 5 min due to road collisions in India, according to an NGO 'Indians for Road Safety'. Speed is the single largest killer on roads in India. According to the data compiled by the 'Ministry of Road Transport and Highways' in 2015, 44.2% (64,633 out of 1above46,133 deaths) of the deaths due to road accidents were a direct consequence of over speeding, while of the total accidents 47.9% (240,463 out of 501,423 accidents) were linked to this. All we have

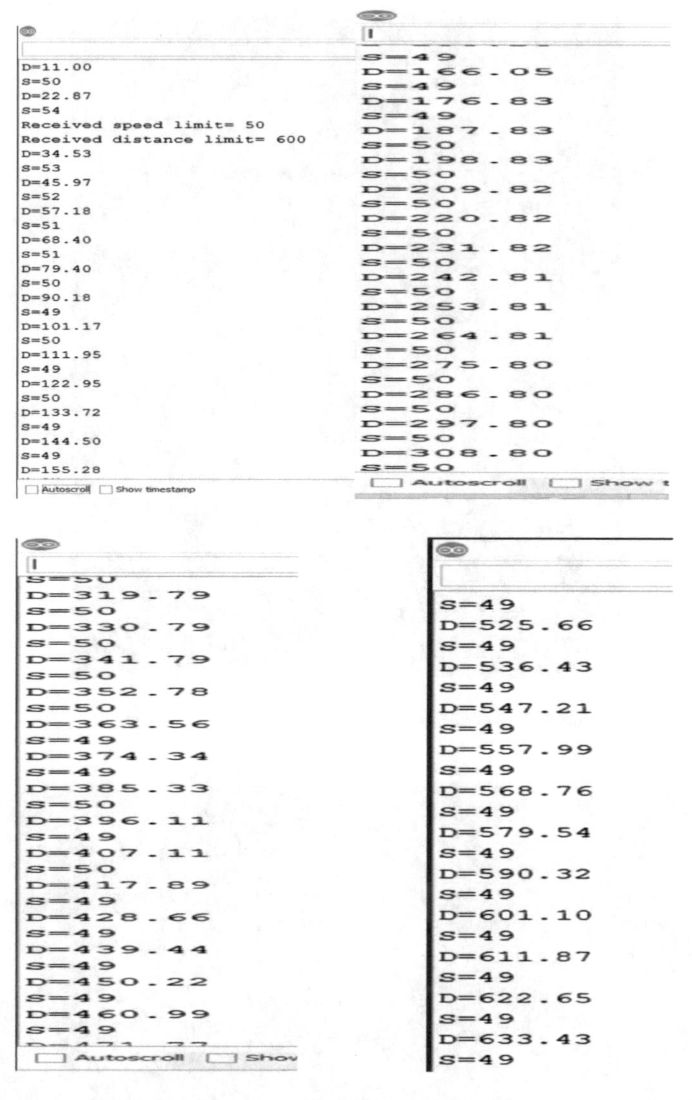

Fig. 12 SCU receiving the speed and distance limits and reducing the current speed to speed limit

to remember is that higher the speed of a vehicle, the larger will be the impact and higher the possibility of dreadful injury or death.

This paper has examined the current solutions for preventing road accidents and has come up with an optimized hardware and software.

A cost-effective system is built that can be installed in electric vehicles without involving much effort.

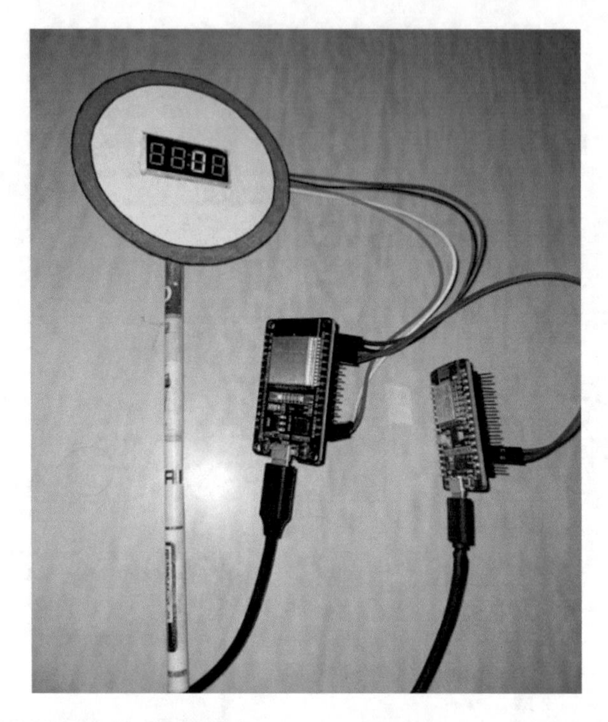

Fig. 13 Road Side Unit with digital display

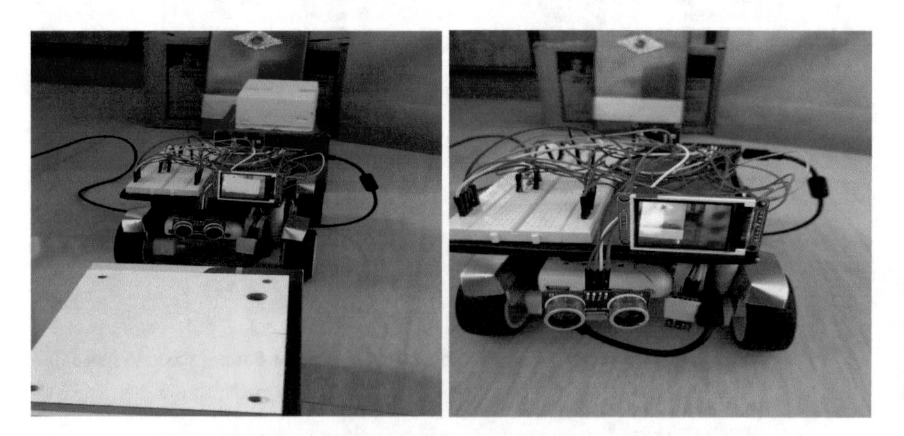

Fig. 14 Video streaming turned on when there's a vehicle at the rear end

Comparing with other communication technologies existing now, BLE is highly energy efficient and has better features compared to others, which led to including the same in the project.

With the see through vehicle feature, i.e., safe overtaking, another cause of road accident is also be reduced.

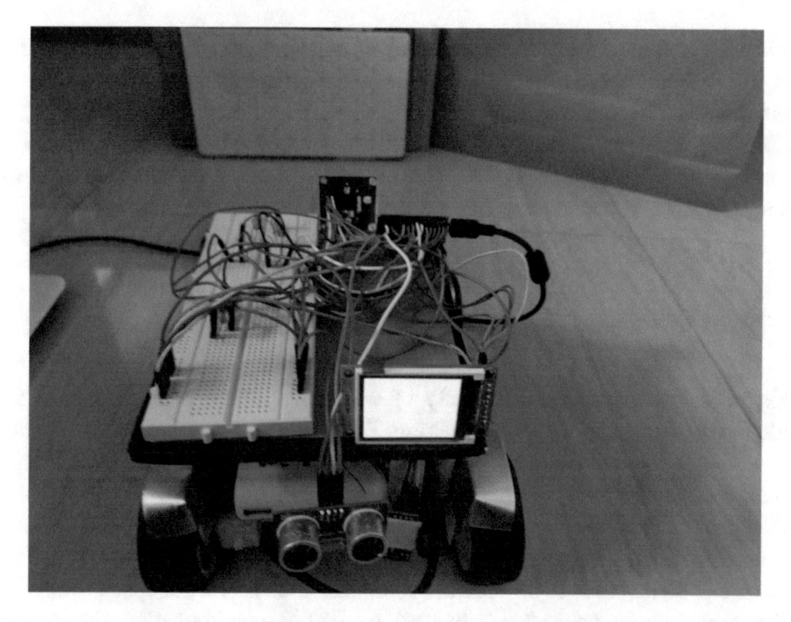

Fig. 15 Video streaming turned off when there's no vehicle at the rear end

Fig. 16 Web-server displaying the speed limit

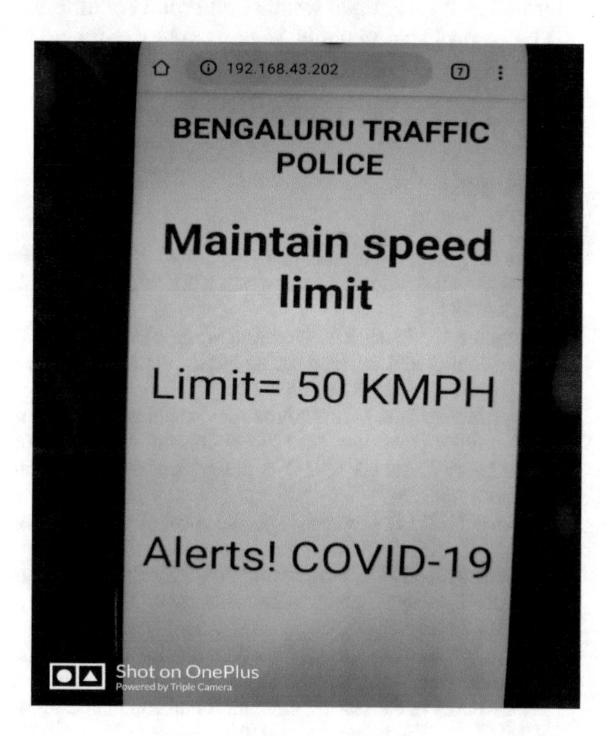

It is strongly recommended that researchers and automobile manufacturers implement this technology as this will help in overcoming road accidents due to over speeding and overtaking.

Companies such as Tesla and BMW are shifting focus from ICEs to EVs and thus influencing automotive electronics. During this shift, novel technologies that make EVs reliable and safe should be implemented.

10 Future Scope

In this project, we are considering only a few sets of the most probable causes for an accident like over speeding and overtaking. The scope for improvements is in making the road network safer while taking other factors like driving under the influence of alcohol, distracted driving, reckless driving, et cetera into consideration.

This project deal needs continuous internet access to connect to the cloud server. This is difficult in rural regions of India. Better Internet facilities or other novel methods can be invented to be used to access the cloud server for speed limit values.

The sign boards could be enhanced so that it can transmit other types of information like regional weather and road conditions.

The control station could be replaced by software which takes all the inputs and predicts the speed values.

References

1. Jeon KE, She J, Soonsawad P, Ng PC (2018) BLE Beacons for Internet of Things applications: survey, challenges, and opportunities. IEEE Internet Things J 5(2). https://doi.org/10.1109/jiot.2017.2788449
2. Prakash SA, Mohan RA, Warrier RM, Krishna RA, Sooraj BA (2018) Real time automatic speed control unit for vehicles. In: 2018 2nd international conference on I-SMAC. https://doi.org/10.1109/i-smac.2018.8653721
3. Rao V, Kumar S (2014) Smart zone based vehicle speed control using RF and obstacle detection and accident prevention. Int J Emerg Technol Adv Eng 4(3):756–777
4. Kameswari J, Satwik (2011) A design model for automatic vehicle speed controller. Int J Comput Appl 35(9)
5. Thomas L (2014) Automatic speed control of vehicles using RFID. Int J Eng Innov Technol 3(11):118–120
6. Satyanarayana KNV, Yaswanthini G, Kartheeka PL, Rajkumar N, BhimaRaju A (2018) IoT based vehicle speed control automatically in restricted areas using RFID. Int J Eng Technol 7(3.31):72–74
7. Volam PK, Kamath AR, Bagi SS (2014) A system and method for transmission of traffic sign board information to vehicles and relevance determination. In: 2014 international conference on advances in electronics, computers and communications (ICAECC), Date of Conference: 10–11 Oct 2014. https://doi.org/10.1109/icaecc.2014.7002415
8. Vengadesh A, Sekar K (2005) Automatic speed control of vehicle in restricted areas using RF and GSM. IRJET 2(9). e-ISSN 2395-0056

9. Anushya D (2018) Vehicle monitoring for traffic violation using V2I communication. In: 2018 second international conference on intelligent computing and control systems. https://doi.org/10.1109/iccons.2018.866300
10. Rothkrantz L (2019) Smart roads. IEEE conference. In: 2019 Smart City Symposium Prague (SCSP). https://doi.org/10.1109/scsp.2019.8805720
11. Rondon R, Gidlund M, Landernas K (2017) Evaluating bluetooth low energy sustainability for time-critical industrial IoT applications. Int J Wireless Inf Networks 24:278-290. 23 May 2017 (Springer)
12. Le-Dang Q, Le-Noc T (2018) Internet of Things(IoT) infrastructures for smart cities. Springer Nature Switzerland AG 2018. 10.1007/978-3-319-97271-8_1
13. Osman K, Ghommam J, Saad M (2019) Guidance based lane-changing control in high-speed vehicle for overtaking maneuver. Springer Nature B V, 25 July 2019. 10.1007/s10846-019-01070-6
14. Mishra A, Solanki J (2012) Design of RF based speed control system for vehicles. Int J Adv Res Comput Commun Eng 1(8)
15. Rao V, Kuma S (2014) Smart zone based vehicle speed control using RF and obstacle detection and accident prevention. Int J Emerg Technol Adv Eng 4(3)
16. Sattibabu G, Satyanarayan (2014) Automatic vehicle speed control with wireless in-vehicle road sign delivery system using ARM 7. Int J Technol Enhancements Emerg Eng Res 2(8)
17. Chavan DB, Makandar AR (2014) Automatic vehicle speed reduction system using Rf technology. Int J Eng Res Appl 4(4)
18. Ankita M, Jyoti S (2012) Design of RF based speed control system for vehicles. Int J Adv Res Comput Commun Eng 1(8):583–586
19. Sattibabu G, Satyanarayan (2014) Automatic vehicle speed control with wireless in-vehicle road sign delivery system using ARM 7. Int J Technol Enhancements Emerg Eng Res 2(8):32–34
20. Chavan DB, Makandar AR (2014) Automatic vehicle speed reduction system using RF technology. Int J Eng Res Appl 4(4):13–16
21. Anitha A (2017) A IOP conference series: materials science and engineering, vol 263, 042027. https://doi.org/10.1088/1757-899x/263/4/042027

Synergy of Internet of Things with Cloud, Artificial Intelligence and Blockchain for Empowering Autonomous Vehicles

C. Muralidharan, Y. Mohamed Sirajudeen, and R. Anitha

Abstract Out of the advancements in Information Technology, the Internet of Things (IoT) plays an important and major role as it metamorphose the object from real world Scenario to intelligent virtual form. The term Internet of Things is coined from two phrases such as Internet and Thing which states that the physical objects such as computing devices acts through the network of connection. This technological ecosystem allows the object or thing to collect and transfer the data through the internet, without any physical assistance. It includes four major processes such as collect, communicate, analyse and act. The main purpose of IoT is making the human life smart thereby reducing the human effort. The cloud is an environment that seems to be a reinforcement of booming technology. It provides everything as service, right from storage to computing power through internet. It seems to be a flexible computing model that has intensified the growth of information technology. It enchanted the sprout of IoT as it needs more storage for the data that are acquired from the objects. Another booming technology is artificial intelligence where the intelligence of machine is used for enabling smart tasks than using the human intelligence. It is in existence since 1950 s but the resurgence of it happens during twenty-first century with the advances in computing power and storage of voluminous data. The main purpose of AI is to achieve accurate interpretation of voluminous data and extract valuable learning from the data thereby achieving the appropriate goals in a flexible manner. The IoT with this gleaming AI allows the physical objects to collect the valuable data through continuous streaming and allows it to perceive its tasks and domains for greatest chance of prosperous goal achievement. Blockchain is another revolution of the information technology. The blockchain or Distributed Ledger Technology is a promising technology where the digital assets of myriad users

C. Muralidharan (✉) · Y. Mohamed Sirajudeen · R. Anitha
DST Cloud Research Laboratory, Department of Computer Science and Engineering, Sri Venkateswara College of Engineering, Sriperumbudur, Tamilnadu, India
e-mail: murali20infotech@gmail.com

Y. Mohamed Sirajudeen
e-mail: ducksirajsmilz@gmail.com

R. Anitha
e-mail: ranitha@svce.ac.in

K. R. Ahmed et al. (eds.), *Deep Learning and Big Data for Intelligent Transportation*, Studies in Computational Intelligence 945,
https://doi.org/10.1007/978-3-030-65661-4_11

are managed by maintaining the transparency and evading the undesirable alterations. It stores and manages the data in the form of multiple blocks with respective cryptographic hashing. It is a distributed and decentralized model where the digital form of transactions are recorded in multiple devices, this allows the system to do any alterations or changes in each and every blocks so as to make changes in the record. This model avoids the precarious changes that may occur in the digital world. The IoT with this blockchain technology or the blockchain of thing may allow the digital environment to create a permanent, verifiable and secure method of managing the valuable data through intelligent machines. It will enable humanless interventions for decision making through proper environment interactions. This chapter elaborates all the four technologies such as IoT, AI, Cloud and Blockchain with regard to the autonomous vehicles. The need for these flickering technologies are explored and exposed so as to understand these technologies. The synergies of IoT with other three technologies are discussed for better understanding and upgradation of the technology. It also scrutinizes the recent developments with all these technological synergies.

Keywords Internet of thing · Blockchain · Cloud · Artificial intelligence

1 Introduction

The internet is a glorious thing that recreates the new world through marvelous inventions that makes the human process easier as it provide us all incredible benefits that were not possible in the past decades. The people who are old enough could think about the cellphones that are used before the emergence of the smartphones. The options that are existed for the mobile users are that they can make a call and can send text to the other users. But after the emergence of the smartphones, the user can even read a book from the place where he is, they can listen to the music, they can send any type of data to others, they access the bank accounts and many, all these processes can be done with the handy mobile. Hence the smartphones are incredible things that made our lives easier. Other devices that support all the above processes are Personal Computers, Laptops, Tablets and so on. All the above processes are possible due to the emergence of the internet.

The Internet of Things (IoT) is a simple system that allows all the things that are existing the world to get connected to the internet. It refers to the model where more number of physical devices that exists around the world are possibly connected to the internet and allows the system to collect and share the data as per our need. The enchantment of IoT is because of the emergence of the super computer chips that are available at lowest price and due to the ubiquitous wireless network. Right from a small pill to the big machineries, the IoT is imposed. All these different objects are connected with each and the sensors are embedded into them that enhances the digital intelligence of the devices and are allowed them to communicate with the real-time data without any human interventions. The Internet of Things makes the system around us more responsive as well as smarter by merging the physical and digital environments. The combination of IoT with other technologies such as Cloud, artificial intelligence and blockchain empowers the field through enhanced performances. This chapter discuss about the synergy of IoT with Cloud, Artificial Intelligence and Blockchain and its applications.

The organization of the chapter is as follows, the Sect. 2 discuss about the Internet of Things, the Sect. 3 briefly explains about the synergy of IoT with cloud, artificial intelligence and blockchain, the Sect. 4 explains about the synergy of IoT with cloud, the Sect. 5 discuss about the synergy of IoT with artificial intelligence, the Sect. 6 discuss about the synergy of Iot with blockchain and Sect. 7 concludes the chapter.

2 IoT

The emergence of the concept of Internet of Things and the term of IoT held during the 1999. It was coined by Kevin Aston who was personnel of Procter and Gamble Corporation. During the Meeting that was handled by the Management of P&G, the idea of radio frequency tags are discussed by the Kevin Aston with the Management which could completely modified the supply chain management system. According to forecasts that are provided by the various analysts, with the increase in the concept of connected devices, "Internet of things" will enlarge around the world. Nearly fifteen years after the emergence of IoT concept, it become the most popular among the existing high-tech technologies. Nowadays everyday objects are connected with each other such as home appliances, thermostats, vehicles etc to the internet via embedded devices, this enables the communication between the things, people and processes. This section discuss more about the internet of things.

2.1 What Is IoT

The Internet of Things (IoT) is the network of physical objects or "things" that are embedded with software, electronics, sensors, and network connectivity, which enables these things to collect and exchange data. In the context of Internet of things (IoT), the "thing" is a physical object or an entity provided with an Unique identifier that has the ability to transfer data over a internet. Few applications where the IoT is used are,

- Heart monitoring implants
- Autonomous vehicles
- Biochip transponders on animals
- DNA analysis devices
- House Automation and other Wearables etc.

All these devices collect useful data with the help of various existing technologies and allows the data to flow between the other involved devices. Though the idea of IoT is in existence of long time, many of the advancements during the current period made the IoT concept to the enhanced practical mode. Few are the advances that enlightened the IoT,

- **Internet Connectivity**: the network protocols of the internet made the sensors to get connected to the cloud environment or to other things for transferring and processing the data.
- **Low-cost and low-power sensor technology**: Reliable and affordable sensors makes the IoT possible for the manufactures.
- **Cloud Computing Environment**: Increased availability of cloud environment enchants both the consumers and the business for accessing scalable infrastructure that reduce the managerial works.
- **Analytics**: The advancement in the analytics enables varied access to the data that are stored in cloud and other platforms with faster and easier manner. This field enhances the boundaries of IoT so that the data produced by the IoT can be limitless.
- **Artificial Intelligence**: the natural-language processing (NLP) is enabled in IoT devices due to the advancements in neural networks that can intensify the automation.

2.2 Process Involved in IoT

The lifecycle of IoT involves several important process where the data is processed over the internet. It includes,

- Collect
- Communicate
- Analyze
- Act.

2.2.1 Collect

The Collect is the first phase where the data is collected by various devices or objects or things with the help of embedded sensors. Eg. Driverless cars, these are autonomous vehicles have many embedded sensors that collects several informations such as speed, direction, location information, traffic information, routing etc. Few other examples where the data are collected from are, home appliances, manufacturing plants, personal devices and so on.

2.2.2 Communicate

The communication is the second phase where the collected data from the sensors are need to be sent to the destination through the network for storing or acting on it. Eg. To make the driverless car, all the events of driver supported car are collected and are stored in the storage medium such as cloud platform, private data center for acting on it.

2.2.3 Analyze

The Analyze is the third phase where the data that is stored in the destination need to be inspected, preprocessed, modified and modeled for retrieving the useful information for achieving the goal. Eg. To make a driverless car, all the events of the driver of the human supported car are retrieved and stored in the storage platform. These data stored are need to be analysed for retrieving the activities of the driver i.e. driving style, braking etc. for imposing it in the driverless car.

2.2.4 Act

The Acting is the fourth phase where the useful information retrieved from the analysed data is implemented to achieve the goal. Eg. Communicating with the other vehicles (Vehicle to Vehicle), Sending the notifications via SMS, Email etc, Sharing Traffic Information, Theft Alert etc. All these processes are happens through wireless internet connection between the connected devices.

2.3 *Applications of IoT*

The automobile sector is one of the biggest and important manufacturing field around the world as the annual production of the automobiles are seems to be around 70 million manufacturing units. An international estimate states that as per the year 2017, the turnover of the industry seems to be around 3 trillion dollars with the global GDP of 3.65%. In recent days, the automotive industry seems to be the important industry where IoT is used with varied enhancements. Since the buyers of vehicles expect futuristic options with which they can interact through the connections. Few are the applications that are related to the autonomous automotive vehicles.

2.3.1 Connected Cars

The connected cars are the cars that are connected through an IoT network that is commonly called as CV2X (Cellular Vehicle to Everything) [1, 2]. It connects the vehicles with the smart transport systems and acts upon through the internet. They are facilitated with fast transmission of data and response time of the drivers are reduced through enhanced communication inside the vehicle [3]. A report by Gartner says that by the end of 2020, around 250 million connected cars will be in use. The CV2X is further subdivided into four type based on the connection with the different objects. It includes,

- **Vehicle to vehicle (V2V)**: This model allows to connect one vehicle with other so as to share data.
- **Vehicle to infrastructure (V2I)**: It enables connection between the network of vehicles and infrastructures of road such as traffic lights, tool booths, lane markings etc.
- **Vehicle to pedestrians (V2P)**: with the V2P the pedestrian get connected to vehicles theerby locating nearby taxis and can estimate the expected time of transition.
- **Vehicle to network (V2N)**: the vehicles are connected to the network i.e. the weather forecast department can alert the drivers of the vehicles through intelligent management system.

2.3.2 Fleet Management

Nowadays the trucks are integrated with location tracking, weight measurement and other sensors for managing the Enlarged vehicle. The large volume of sensor data collected from the trucks can be stored in the cloud and are analysed for extracting the important features by making it to a visualized form. The operator of the fleet can view the information in the monitor. Some benefits of IoT imposed fleet management system are

- Real-time monitoring of the fleet location
- Cargo Weight or Volume tracking
- Traffic conditions monitoring
- Statistics of Trucks' performance such as mileage and fuel availability
- Route management etc.

2.3.3 In-Vehicle Infotainment

The vehicle owners are able to interact with their vehicles through smart enables dashboard through the security, safety and surveillance of the vehicle can be ensured at all the times. The External sensors keep a track of the condition of the vehicles and send the data to the user's application. The smart Telematics system that exists in the vehicle with the real-time alert system awake the owner's of the vehicle through their smartphone if someone tries to enter the vehicle forcefully without any proper access.

2.3.4 Automotive Management and Maintenance System

Predictive analytics is an important feature of the automotive IoT. The sensors that are embedded in different parts of the car collect the data and store it in the desired platform, later this data is been analysed using the analytics algorithm for extracting the future outcomes of the parts of the car based on the collected performance. Like the smart dashboard of the vehicle, this system alerts the owner or driver to get awareness about the malfunctions in the vehicle thereby reducing the sudden breakdown.

2.4 Challenges of IoT in Autonomous Vehicles

The Autonomous vehicles are already in use on our roads we are sure about the concept that whether the drivers are really ready for the self-driving cars [1]. Few are the some technological challenges that need to overcome.

- **The Data Traffic**: The large amount of data is been collected constantly by the available high quality sensors. This type of information is essential for stakeholders such as insurers, manufactures, emergency services, councils etc. Though the idea is good, interacting with all the stakeholders make the data traffic heavy.
- **Data Security**: It is an important field that is to ensure the security of the vehicle data. As the smart system enables to share the data with different stakeholders, it is not sure about how it is transferred between the communities whether it is to be shared in an open manner or through proper trusted parties [4]. Also Plenty of hackers are emerging in the internet who are found to be threat for the vehicle [3].
- **Cope up with Traffic**: Though most of the vehicles are tested and racked up in use on highways, the self-driving cars seems to have lesser experience to deal with the disordered conditions of the urban traffic. still there is not any assured practical experience for ensuring cyclists, children safety, cyclists and older people who use the road. Also to deal with the busy and stop and go traffic interaction scenarios, the vehicle needs an V2X connection with strong bandwidth of network.
- **Crowd Navigation**: There will not be any serious issues with the road crossing people in the developed countries but in the crowded cities the autonomous vehicles face the problems with the crossing [3]. The autonomous car will wait though the signal turns to green when group of people are waiting to cross the road in the zebra crossing. Though the situation works good, the traffic will get increased in busy places like Mumbai.

3 Synergy of IoT with Other Technology

The Internet of Things is a technology that interacts with other emerging technologies for various purposes. In this chapter the interaction of IoT with other in the field of autonomous vehicle is discussed. Below are the few synergy of IoT with other technologies,

- Cloud Centric IOT
- AI centric IOT
- Blockchain Centric IOT.

3.1 Cloud Centric IoT

Cloud is a platform where everything is provided as service such as Infrastructure as a service, Platform as a service and so on. It becomes daily part of our lives as the use of online data processing tends to increase. The main purpose is Infrastructure which reduces the maintenance of the organization which uses the cloud [5, 6]. The user can store any data such as photos, personal information etc. and can be accessed over the internet from where we are. This lead the enhancement of Internet of Things platform as it supports the IoT devices by providing the storage platform for it [6]. It also provides analyzing platform through which the data stored in the cloud can be retrieved and analyzed in the cloud itself. The Autonomous vehicle have more number of sensors with it and all these sensor collect the data from different parts of the vehicle, that will stored in the cloud for further process [2, 5]. With the cloud platform, the vehicle that run on the road communicate with other vehicles through which the accidents can be avoided, information about the traffic can be retrieved, quick route can be extracted etc. Hence the cloud plays a vital role with IoT. This section is further discussed deeper in Sect. 4.

3.2 AI Centric IoT

The Artificial Intelligence (AI) is a tool for abstracting the intelligence of machines thereby enabling the revolution in smart industrial technologies. The process flow of artificial intelligences includes several steps such as collecting the related information, identifying the important features, taking necessary actions, decision making, reviewing and finally predicting the future for making the system smarter [7]. On the contrary, the IoT is the field where the data is collected from the sensors and are analyzed for further process. On combining both the technologies we could say that the first process of collecting information or data can be done by using the sensors and are stored in the storage platform. The stored information can be analysed by the Artificial Intelligence platform or tool so as to extract the future predictions for making the system smarter. Both of these platforms could be an axiom for the industry 4.0. With the high speed connectivity, the IoT and the AI transforms the autonomous vehicle into fully smarter vehicles which illustrates the knowledge of real and digital world for the industry 4.0. This synergy of IoT with the AI is discussed more in Sect. 5.

3.3 Blockchain Centric IoT

The blockchain is the list of blocks or records that are linked with each using the cryptography. Each of the blocks will have a cryptographic hash value of the prior block so as to make the connection, a timestamp at every block and the transaction data. It is evolved in recent years and are created in the fame of crypto currencies and now it is been used or applied in various applications including industries [8]. It is expected that in next few years the digital economic transactions of blockchain will be integrated to the vehicles. The primary use of it in the autonomous vehicles is that ledger maintenance with the blockchain concept which uses multiple servers so as to track the networks of the vehicles through integrated sensors in the vehicles [9]. Smart sensors, as used by Adaptive Cruise Control with the blockchain, and the IoT will improve the vehicle's ability to act without the human intervention thereby, enabling smooth movements, keep away from the upcoming traffic in the current route and change the route without needing to tap the brakes manually. The interactions between the IoT and the blockchain is elaborated more in the Sect. 6.

4 Cloud Centric IoT

While observing the nature of Internet of Things (IoT) device in autonomous vehicle, it represents a comprehensive environment which consists of a large number of sensors, interconnecting devices, heterogeneous microcontrollers to sense and communicate the vehicular information to the server [10].

Though, IoT is a promising technology, it also faces several bottlenecks. Firstly, the computation and storage power of IoT devices are very less compared to the PCs and laptops. Performing a heavily load task is not possible in the IoT devices. For example, if an object suddenly falls in front of an autonomous vehicle, then the IoT devices present in the vehicle cannot make a decision to apply break. Instead the information about the object is detected and sent to a server, which has an enormous capacity of computation power. The server will perform artificial intelligence, machine learning, and data analysis to make the decision.

Moreover, the data generated through the vehicular IoT devices are exponentially growing. So, holding the sensed data for a long time in vehicle IoT devices are not possible. The information received from the sensor has to be immediately forward or uploaded to the server [11]. If not, the sensed information could be deleted in a very short period of time due to its low storage capacity.

The second important problem that IoT faces is, secure communication of sensed data. Securely transmitting the IoT data to server is a really a challenging tasks. According to the statistics of HIPPA online journal, a total of 3054 data breaches have happened between the year 2009 and 2019, in which 510 breaches were betided in 2019. As a total of 230,954,151 IoT generated data were leaked, theft, impermissibly disclosed. It also mentions that the number of exposed records has tripled between the year 2018 and 2019. Since, the Vehicular IoT data are used in real time applications, data breach could cause a terrible accidents.

The third most problem with IoT is, data management. According to the recent report of International Data Corporation (IDC), the number of IoT devices connected to the internet will reach 41.6 billion IoT devices. It also mentions that the data generated through these IoT devices will reach 79.4 Zetta Bytes (ZB) in 2025. Managing such a huge amount of data needs a hue or an unlimited capacity of storage.

Precisely, the challenges associated with IoT can be summarized in to three categories, such as (i) Limitation of storage and computation capacity (ii) Secure communication of sensed information (iii) Managing the data generated by the IoT devices.

To overcome these issue, Cloud-Centric IoT architecture is used. With the recent exploration of Internet technologies and cloud computing technologies, both private and public organizations are using cloud service models such as Infrastructure as a Service (IaaS), Platform as a Service (PaaS), Software as a Service (SaaS) and Anything as a Service (XaaS) along with cloud deployment models (public, private and hybrid) to deploy their day-today applications [10]. The main advantage of using cloud services is, it massively reduces the capital cost of setting the infrastructure. The cloud services allow its users to get access to higher-end equipment's in a rental basis. Though, the user has very limited computation and storage powered equipment, cloud allows to work on high-end machines for a very minimal cost. Moreover, the capacity of cloud services are unlimited. Figure 1 explains the basic architecture of cloud centric IoT applications.

In addition to the unlimited storage capacity, other advantages of using Cloud-Centric IoT architectures is virtualization. It is an act of creating a virtual version of a computer application, storage device or a computer network resource (i.e.) It

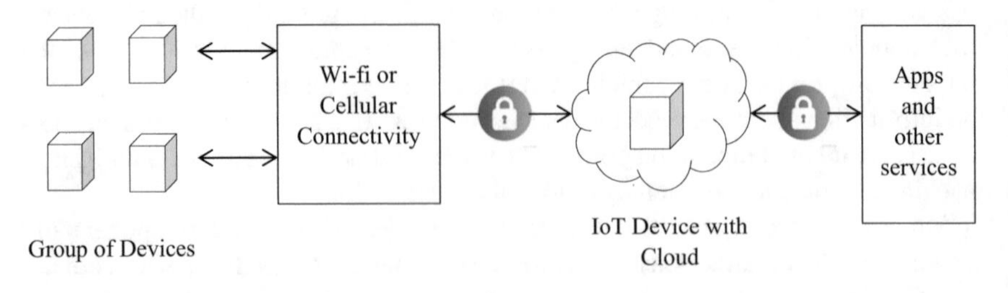

Fig. 1 Cloud centric IoT architecture

makes logical server, storage and network independent of the deployed infrastructure resources. It allows the cloud users to request the exact infrastructure needed to run the user application via internet.

In general virtualization in Cloud centric IoT infrastructure can be categorized into (i) Device Virtualization (ii) Network Virtualization.

4.1 Device Virtualization

Device virtualization in cloud centric IoT architecture allows to virtualize an IoT device in the data plane using a certain logical abstractions among its components. It provides software abstraction of various external IoT devices (developed by different manufacturers) and allows application to easily utilize it through application programming interfaces (API) [10]. It can represent any object on the device side that can send messages or respond to REST requests. Device virtualization includes, includes devices, gateways, device adapters, and device applications. Some of the openAPI, which offers device virtualization are Sphere3D, Glassware 2.0 and Zerconf API. In device virtualization, the virtual sensors communicate through standard communication interfaces such as UDP, TCP or HTTP. Through, device virtualization, user can access, metadata associated with the device, message formats generated by the device, web resources to send commands and software management.

4.2 Network Virtualization

In recent times, two of the most important phenomena of networking are, Internet of Things (IoT) and the virtualization of networks as a Software-Defined Network (SDN) and Network Functions Virtualization (NFV). Network virtualization (NV) combines the network resources such as hosts, adapter, switch, router, virtual machines into a combined platform that appears to be a single pool of service. It is implemented using software containers on a single, physical, host machine to run multiple, virtual, guest machines. Network virtualization increases the resource utilization by sharing network resources among multiple users [10]. Both platform and resource virtualization are needed to form a successful network virtualization. As like other virtualization techniques, NV also needs an addition abstract virtualization layer between the network and storage hardware. In general, NV can be categorized into (i) external virtualization and (ii) internal virtualization. Here, external virtualization consists of many networks into a single virtual network unit and internal virtualization consist of network-like functionality to software containers on a single network server. Using IoT technology with Network virtualization can improve the efficiency, scalability and reduce the operational cost and complexity.

4.3 Applications

The Internet of Things (IoT) refers to the connectivity of multiple devices through the Internet. Driverless cars utilize this connectivity when updating their algorithms based on user data. These autonomous vehicles require an enormous quantity of data collecting and processing [12]. In this case, through IoT, the driverless car shares information about the road (which has already been mapped out). This information includes the actual path, traffic, and how to navigate around any obstacles. All of this data is shared between IoT connected cars and is uploaded wirelessly to a cloud system to be analyzed and put to use improving the automation. While being connected through IoT is speeding up the development of current-day self-driving cars, people have been trying to tackle this idea for years [12]. There are many companies working on a driverless car. Here is a list of some of the most notable and recognizable names: Uber, Honda, Toyota, Tesla, Hyundai, Volvo, Waymo (partnered with Google, Fiat Chrysler), BMW, Volkswagen, General Motors, and Ford.

Amongst all of these working on this technology, the companies that we hear the most about are Tesla, Uber, and Waymo (Google). But regarding driverless cars, the news coverage isn't merely positive. Just in the last couple of months, multiple accidents have been reported involving autonomous vehicles.

5 IoT with AI

Artificial Intelligence is defined as the ability of the system to achieve the specific goal by interpreting the related data correctly [13]. In deeper manner, a device itself will observe the environment, analyse it and makes the decision in the context to achieve the specific goal. Combing artificial intelligence and cloud centric IoT applications has become a new trend in data analytics. Both investments and acquisitions in startups that merge AI and IoT have climbed over the recent years. Major vendors of IoT platform are offering integrated AI capabilities such as machine learning-based analytics as a service to their clients. According to the recent Gartner report, almost 80% of enterprises which uses cloud centric IoT architecture are using AI as it major component for identifying the pattern/prediction. It also states that around 41% of IoT users believes that the AI will ease their lives.

Artificial intelligence and machine learning techniques has an ability to find out the insights of the data and automatically identifies the pattern of the data generated from the IoT devices. Traditional machine technique were using offline data or previously collected data to identify the patterns. But after the emergence of cloud centric IoT application, the data are collected in real time and processing AI and machine learning technique to identify the pattern. The Autonomous cars are increasing in the recent days which dominates the tech-talks. It is seen as the post-Uber disruption for the public commute and for goods transportation. There are no figment of imagination that the AI is used to complement driverless vehicles. AI is considered to be the game changer of autonomous vehicle by enabling driverless vehicle, predictive maintenance, driver behavior observance etc.

5.1 Synergy of IoT and Artificial Intelligence

In this section, some of the AI and IoT combined services such as ET city brain, Tesla Auto Pilot and Remote controlled cargo ships are explained.

5.1.1 ET City Brain

ET city Brain is an IoT based artificial intelligence platform, which was developed by Alibaba Cloud services to help the citizens of Hangzhou, China to optimize the usage of public resources. It collects the data from the public resources such as traffic, accidents, illegal parking and ambulance service in real time and performs artificial intelligence computations to address the day today needs. ET city brain claims that the road traffic in Hangzhou is reduced by 15% in 2019.

5.1.2 Tesla Autopilot

The Tesla autopilot system enables GPS, sonars, cameras and forward-looking radars, in combination with specialized hardware, through which data can be fully utilized and coupled into Neural Network Architectures. This works like a self-enclosed system that gathers information from the sensors and further uses the Neural Network model that determines the next change in the movement of the car.

5.1.3 Remote-Controlled Cargo Ships

The famous automobile brand Rolls-Royce have revealed a plan of making a human-less cargo ships for transporting the goods. The idea behind the model is the cargo ships are to be controlled by a land based control hub and since the space for on board crew is reduced, the ship might have more space for goods. It is partially automated type of ships that could be controlled by a remote control and they were in the plan to make it within a decade.

5.2 Components of AI with the IoT

As per IBM, four major components are considered to the pillars of artificial intelligence. It includes, Faireness, Robustness, Explainability and Lineage [12].

- **Fairness**: The data from the IoT devices should be fair data so that the training data used by the artificial intelligence system will be free from bias which further reduces the unfair groups.
- **Robustness**: The data that are to be processed by the AI system should be secured and should not be tampered or compromised as it will be used for training the system.
- **Explainability**: The decision made by the AI systems has tobe understood by the developers and the users.
- **Lineage**: The details such as development, deployment, and maintenance should be mentioned by the AI system which might help to audit the system throughout the lifecycle.

5.3 Applications of IoT with AI

Below is the few application of IoT with AI in the field of autonomous vehicle.

5.3.1 AI for Self-driving Car Safety

Before implementing the driverless car, the AI system is used as a co-pilot for obtaining the confidence of the users, regulators, manufacturers. Thus system analyzes the data feeds that are captured by its sensors so as to overcome the human errors that might occur in the non-autonomous car. It is also good in the services such as

- Emergency control of the vehicle
- Traffic detection
- Traffic signs synchronization
- Emergency braking
- Active monitoring of blind spots.

5.3.2 Cloud Services Enabled Vehicles

The artificial intelligence system with the sense of IoT accurately gauge the changes in the vehicle, hence it tracks the physical conditions of the vehicle. All these data collected from the usage will be further used for:

- Predictive maintenance
- Prescriptive maintenance.

A car warrenty plan might be enabled which might satisfy the needs o the driver as it updates the condition of the car periodically.

5.3.3 Auto Aviation

Dubai have collaborated with Volocopter for using the drone air taxi which is a two seater humanless vehicle that fly in air. The vision of the project is to introduce an air taxi for reducing the battle with the traffic in roads. Likewise Amazon extending its delivery medium to aerial vehicle called prime air for reducing the delivery time of the product to the customer.

5.4 Advantages of IoT Powered AI

It is considered that the Ai seems to be the future of IoT which has many advantages as,

- Avoids Unwanted/unplanned Downtime
- Increases the operational efficiency
- Enables the improved services and products
- Enhance the risk management
- Implicate the Enterprises.

6 IoT with Blockchain

The blockchain technology has become a new backbone of internet which allows the digital information to be distributed. It is first created for the implementation of bitcoin, digital currency but now it is used for other technologies because of its performance. It is a simple term which includes immutable records of data with respective time stamps that are managed by group of cluster of computers. It is in the form of blocks and it imposes the security over the data by bounding the principles of cryptography [14]. Dan Beiler—an analyst at Forrester states that the blockchain is a ledger technology that will play an important role in the field of internet of things to manage the direct communication of devices with each other. This technology maintains a ledger which tracks not only the trails of devices but also the tracks the interaction of devices, and the potential status of the devices. With the increase in the arrival of connected cars which are embedded with multiple sensors, cloud based infotainments, telematics system etc, the generation of data becomes higher that increases the burden of management and security enrichment. The automotive industry to solve the above problems intend to gain the blockchain technology application. It is expected that the vulnerable and hacks over the data that are collected from the highly featured vehicles can be secured on imposing the blockchain technology with the internet of things [15]. The grounds and application of block chain with the IoT in autonomous vehicles are discussed in the below sections.

6.1 Grounds of Blockchain

During the period of 90's, the Staurt Haber and W. Scott Stornetta started their first work on blockchain by creating the secured chain of blocks by using the cryptograpy. In further research, they tried to secure the timestamps of the document that might be tampered. Later Haber, Stornetta and Dave Bayer tried to improve the efficiency of the concept by using the Merklee trees which allows the model to collect the certificates of the documents in one block. During 2008, the concept blockchain is coined by Satoshi Nakamoto. He used the hashcash methods for the timestamp blocks without the authorization from the trusted party and parameter is introduced for adding the blocks to the chain. This concept has its core idea of crypto currency called bitcoin which usually holds a public ledger for tracking all the transactions in the network. The blockchain is a list of records commonly called as blocks that impose the principles of cryptography for secure transactions of data. It is in the form of several blocks which the blocks holds the prior block cryptographic hash value, a timestamp of the data and the transaction data. It is resistant to data modification and is a digital distributed ledger that holds the traction records that happens between the two or more parties in a permanent and verifiable way. The three main pillars of the block chain technology are decentralization, transparency and immutability.

6.2 Need for Blockchain in IoT

IBM blockchain have extended it process by combining the concept of blockchain and the internet of things. IBM provides the concept of complex trade logistics in which the block chain register can track all the events that happens right from individual item to packages. It is expected that by the end of 2020, there would be millions of connected cars around the world that would be highly featured cars with more number of sensors and other cloud based infotainments and telematics. This might increase the generation of data as well as flow of data. Hence secure channel has to be imposed for both transaction and storing of data that are collected from the vehicles. Experts says that by implementing the Blockchain with the internet of things, the vulnerability of data from the hackers can be reduced. As the information technology field, the manufacturing technologies have essential need for revolving around the huge amount of data such as data capturing, leveraging, interoperable, ecosystem leveraging etc. The plenty of data are need to be exchanged either between the manufacturing plant or to the other mediums, hence there is a need for implementing blockchain for both tracking the transactions and securing it. Also there is a need of making decisions autonomously through proper analysis.

6.3 Blockchain of Things

The automotive industry tend to gain the blockchain technology application because of its performance. Since the upcoming high-tech features are about to increase in the vehicles, the vehicles becomes connected one through network which might generate larger data that need to handled securely as many vulnarabilities and threats can tamper the data [15]. To overcome these issues, the blockchain technology is used along with the IoT as it is holds the encrypted and fully immutable digital database which contains the plenty of information that can be accessed within the decentralized and publicly available networks through proper smart contracts. These smart contracts highlights the information about the usage of data i.e. what or who can access the data. It is applied to the connected cars for many purposes as,

- Vehicle safety and data security
- Transparency in Supply chain
- Automated Financing
- Infotainment and Telematics
- Manufacturing
- Smart Contracts
- Fleet Management.

With the combination of blockchain technology and 5th Generation to the automated vehicles, important processes such as monitoring, protection from threats, tracking and information sharing etc can be possible. Since it enables resource pooling and transaction verification, it can be leveraged for the services such as vehicle platooning. This service allows the group of vehicle to negotiate between them and agree to form a group or chain between the smart enabled cars or trucks that are in the minimal distances [16]. This can possible have it benefits over fuel efficiency, lower commute times and collision avoidance.

6.4 Applications of Blockchain with IoT in Autonomous Vehicles

Though the blockchain technology is primarily used for tracking the networks where the crypto currencies are shared, it has more use in other evolving technologies. Especially in the field of automotive industry, it is intensely used for monitoring the transactions that happens with every vehicles that are imposed to smart system. Few application of the usage of Blockchain with the IoT in the autonomous vehicles are as follows,

6.4.1 Interconnectivity Improvements

The shift towards the automated battery cars has made the auto engineering more creative. This emphasis on the enforcement of automation uses the Internet of Things for finding the home from your car. The car can be connected to the owners phone through the mobile application and are allowed to use their favorite playlists in the car's sound system [15]. The integration of blockchain with this IoT devices expand the possibilities in the digital world. The autonomous vehicles has the adaptive cruise control options and with the combination of blockchain with the IoT device that handles the cruise control will improve the vehicle ability to work without the human intervention and can offer smooth movements to the vehicle during the traffic by avoiding the taping of brakes.

6.4.2 Easy Transactions

In terms of economics, it is considered that many numbers of transactions will take place while we are in the road. For example, Toll Plaza where much time is needed for the required transactions and potentially cause unwanted traffic. On the other hand the hacker can steal the card details from the toll plazas by themselves. On integration of blockchain into the smart vehicle, the transactions can be made private. For example, Car e-Wallets where the imposed blockchain will secure the transactions by keeping the information of the users as private.

6.4.3 Ride Sharing

There is much increase in the rise of the ride-sharing industry. Over the past decade, Uber and the Lyft have made a shift in the automobile industry due to the service they offered to the people. The ride-sharing, a billion dollar company offers blockchain and e-commerce the opportunity to thrive. This technology allows the user to make payment without the use of credit card and will support the driver who fails to collect the payment from the loiter. The ride sharing companies uses autonomous driving which tracks the every transactions or ride details for making the system interaction free.

6.4.4 Personalized Car Insurance

The insurance agencies are in the need for the use of blockchain. For example, The Mobility Open Blockchain Initiative, who works with Ford and GM for instituting the user-based insurance. Once the Insurance contracts are established, it will be stored in the blockchain servers. Once the sensor in the car sense the payment contract violation, the payment can be made automatically. Service such as telematics collects the vehicle driver's road history which covers location, drive length and speed of driving so as to report to the insurance company for determining the premium mode.

7 Conclusion

This chapter discussed all the four technologies such as IoT, AI, Cloud and Blockchain with regard to the autonomous vehicles. The need for these flickering technologies are explored and exposed so as to understand these technologies. The synergy of IoT with other three technologies are discussed for better understanding and upgradation of the technology. The recent developments with all these technological synergies are also discussed for better understanding.

References

1. Rosenzweig J, Bartl M (2015) A review and analysis of literature on autonomous driving. In: The Making of innovation, pp 1–57
2. Shanker R et al (2013) Autonomous cars: self- driving the new auto industry paradigm. Morgan Stanley Blue Paper
3. Joy J, Rabsatt V, Gerla M (2018) Internet of vehicles: enabling safe, secure, and private vehicular crowdsourcing. Internet Technol Lett 1(1):1–16
4. Alam KM, Saini M, El Saddik A (2015) Toward social internet of vehicles: Concept, architecture, and applications. IEEE Access 3:343–357
5. Gerla M, Lee E-K, Pau G, Lee U (2014) Internet of vehicles: From intelligent grid to autonomous cars and vehicular clouds. In: 2014 IEEE world forum on Internet of Things, WF-IoT 2014, pp 241–246. https://doi.org/10.1109/WF-IoT.2014.6803166
6. Silberg G (2012) Self-driving cars: the next revolution. KPMG LLP and the Center for Automotive Research
7. Khayyam H, Javadi B, Jalili M, Jazar R (2020) Artificial intelligence and Internet of Things for autonomous vehicles. https://doi.org/10.1007/978-3-030-18963-1_2
8. Rathee G, Sharma A, Iqbal R, Aloqaily M, Jaglan N, Kumar R (2019) A blockchain framework for securing connected and autonomous vehicles. Sensors 19(14):1–15
9. Boudguiga A et al (2017) Towards better availability and accountability for IoT updates by means of a blockchain. In: 2017 IEEE European symposium on security and privacy workshops (EuroS&PW), Paris, pp 50–58. https://doi.org/10.1109/eurospw.2017.50
10. Celesti A, Galletta A, Carnevale L, Fazio M, Ĺay-Ekuakille A, Villari M (2018) An IoT cloud system for traffic monitoring and vehicular accidents prevention based on mobile sensor data processing. IEEE Sens J 18(12):4795–4802. https://doi.org/10.1109/JSEN.2017.2777786
11. Datta SK, Da Costa RPF, Härri J, Bonnet C (2016) Integrating connected vehicles in Internet of Things ecosystems: challenges and solutions. In: 2016 IEEE 17th international symposium on a world of wireless, mobile and multimedia networks (WoWMoM), Coimbra, pp 1–6. https://doi.org/10.1109/wowmom.2016.7523574
12. Shahzad K (2016) Cloud robotics and autonomous vehicles, book on autonomous vehicle. Intech Open

13. Nanda A, Puthal D, Rodrigues JJPC, Kozlov SA (2019) Internet of autonomous vehicles communications security: overview, issues, and directions. IEEE Wirel Commun 26(4):60–65. https://doi.org/10.1109/MWC.2019.1800503
14. Miller D (2018) Blockchain and the Internet of Things in the industrial sector. IT Prof 20(3):15–18. https://doi.org/10.1109/MITP.2018.032501742
15. Singh M, Singh A, Kim S (2018) Blockchain: a game changer for securing IoT data. In: 2018 IEEE 4th world forum on Internet of Things (WF-IoT), pp 51–55. https://doi.org/10.1109/WF-IoT.2018.8355182
16. Pustišek M, Kos A, Sedlar U (2016) Blockchain based autonomous selection of electric vehicle charging station. In: 2016 international conference on identification, information and knowledge in the Internet of Things (IIKI), Beijing, pp 217–222. https://doi.org/10.1109/iiki.2016.60

Combining Artificial Intelligence with Robotic Process Automation—An Intelligent Automation Approach

Nishant Jha, Deepak Prashar, and Amandeep Nagpal

Abstract Process Automation has the potential to bring great benefits for businesses and organizations especially in the financial services industry where businesses are information-intensive and experience rich data flows. This was achieved mainly via Robotic Process Automation (RPA), but the increased complexity of the Machine Learning (ML) algorithms increased the possibility of integrating classic RPA with Artificial Intelligence (AI), leading to Robotics 2.0. However, the transition from RPA to Robotics 2.0 embeds a number of challenges. To ensure that the advantages of the modern technologies can be harnessed, these issues need to be tackled. By integrating RPA with cognitive technology such as machine learning, speech recognition, and natural language processing, businesses can automate higher-order tasks with AI assisting that human perceptual and judgment skills were needed in the past. The purpose of this chapter is to identify the set of challenges the companies will face, as well as provide guidance on what preparations to be made before Robotics 2.0 can be implemented in full scale. This also provides the insights about the new intelligent automation approach based on AI integration with RPA in intelligent transportation system.

Keyword RPA · Robotics 2.0 · Machine learning · Artificial intelligence · Cognitive technology · Process automation · Financial services

N. Jha · D. Prashar (✉) · A. Nagpal
Department of Computer Science, Lovely Professional University, Punjab, India
e-mail: Deepak.prashar@lpu.co.in

N. Jha
e-mail: nishant.11702196@lpu.in

A. Nagpal
e-mail: amandeep.nagpal@lpu.co.in

K. R. Ahmed et al. (eds.), *Deep Learning and Big Data for Intelligent Transportation*, Studies in Computational Intelligence 945,
https://doi.org/10.1007/978-3-030-65661-4_12

1 Introduction

Artificial Intelligence is the field of science that enables computers and machines to understand, judge and manipulate their own reasons. As the technologies become more complex, the demand for Artificial Intelligence is increasing due to its ability to solve complex problems with limited human intervention and expertise and within a short period of time. AI takes the opportunity to develop technological skills and improve awareness for the development and delivery of new approaches and applications. There is a major advancement in the field of image recognition, using machine learning along with developments in big data and GPU (Graphic Processing Units), which naturally helped Artificial Intelligence to grow faster than other fields. By integrating RPA with cognitive techniques such as machine learning, speech recognition, and natural language processing, businesses can automate higher-order tasks with AI assisting that human perceptual and judgment skills were needed in the past. Over the last decade, the use of automation started to increase, with the goal of reducing manpower and time. Automation implemented a computer and machine system, which replaced a system designed by combining man and machine. Through the use of automation in various industries, highly intense and repetitive activities have become productive, and the product quality has also increased. In many industries automation with time is becoming a go-term. The automation has become omnipresent, from self-driving vehicles to social media messages. According to a report by The Gartner [1], Robotic Process Automation (RPA) software sales grew 63.1% in 2018 to $846 million. This is the world's fastest-growing tech enterprise market. This is also expected to hit sales of $1.3 billion by the end of 2019. Although RPA today is entering almost every sector, banks, insurance companies, telecom firms and utility companies are the biggest adopters of this tech. This is because businesses typically have outdated structures in these industries, and RPA solutions are easily compatible with their current functionalities. RPA is often referred to in the same breath as artificial intelligence, deep learning, machine learning and the processing of natural languages. There are variations however; most people believe that any part of automation is artificial intelligence, which is not valid. RPA and AI are two horizontal systems that have separate aims and configurations.

RPA is software that automates computer programs to perform tasks in compliance with a set of rules in line with the business process. With the introduction of emerging technologies within Artificial Intelligence (AI), RPA can also be further established by learning from experience to carry out tasks that involve human cognition. The introduction of these methods has the ability to fully eliminate many fronts, middle and back-office functions. Several challenges emerge with developments like these, both in terms of technical viability and human capital. This area of research is very young and several organizations have yet to adopt RPA from an implementation point of view. This makes it especially important to disguise the path to the next level of intelligent robotics. AI and Machine Learning (ML) are among this decade's most influential buzz-words. The ideas have at least been learned by almost every modified firm in all industries. Many of the organizations are, however, far from actually using it in their organization, and one explanation for this is that the partnership with AI still does not go beyond learning about it. In order to make efficient use of the technology, and not to waste money and time trying to implement something that only sounds amazing conceptually, it is vital to create a suitable knowledge base from which effective implementation of AI will arise [2].

Through cloud-based ML services, major technology companies such as Google, IBM, and Amazon have come a long way to develop open AI technology, making the technology more affordable. While some of the technology is in place, a huge portion of the work still remains to be done for organizations that want to adopt the features, as technical developments within AI are on the verge of a fully digital market disruption [3]. Check out Face book's AI Work to get a better understanding. Here the social media giant feeds different images into the AI system, and the computer produces exact results. When the computer is shown a picture of a dog, it not only recognizes it as a dog, but also the breed. RPA is a technology that uses a specific set of rules and an algorithm and automates a function based on that. Although AI is more focused on performing a human-level job, RPA is basically a machine that eliminates human effort—it's about saving time for the company and white-collar employees. While AI is steps ahead of RPA, if both are combined, these two techs have the potential to carry things to the next level. For instance, suppose you need your documents to be scanned in a specific format, and RPA does that job. If you are using an AI program which would filter out badly written or unacceptable documents, the RPA's work will be much simpler. And this relationship is called Continuum Automation or Robotics 2.0.

2 Implementation Technique of RPA + AI

We have a user interface starting from the top which will allow user requests to be captured. The user's purpose is understood by classifying NLU pipelines into predefined categories. These pipelines usually consist of many NLP libraries such as SpaCy or NLTK, offering multi-language support for word vectors/embedding tokenization, etc. The essential function that this layer often performs is to extract user requests for the entity (contextual data) type. Entities are part of a text of interest to the data scientist or the company, such as people's names, addresses, account numbers, locations etc. These are then moved downstream to be processed further. Advanced techniques and algorithms, such as CRF (Conditional Random Fields), Stemming, etc. are used to train Named Entity Recognition (NER) models to perform. NER support is available in libraries such as SpaCy which allows the extraction of entities by knowing their pattern and other statistical properties. In addition, basic pieces of information can be collected using libraries such as duckling, such as date and time. When the intent and the objects of the data are extracted, they pass through the next level, which is the interface Machine to Machine. A software robot may be plugged in at this stage to drive aspects of orchestration and (mainly) automation. For instance, 'Create and send a report,' 'provide or set up a network device' or 'use a PayPal account to buy movie tickets'. Figure 2 will excessively simplify efforts to incorporate the architecture shown in Fig. 1. There are many moving parts here, and for successful implementation and coordination, each portion (sub-system) needs considerable effort and proper domain knowledge. In the design process, it is prudent to take a deep holistic look and hammer out details such as protocols, data structure, access specifications, performance and quality, availability, tracking and monitoring, managing exceptions, as well as leveraging technology. Externalizing and decoupling the control logic, where possible, will offer benefits in terms of versatile design and device updates.

2.1 Exception Handling in the Combination of RPA + AI

There will be cases where one-off situations such as 'insufficient balance to buy grocery,' 'invalid shipping address,' 'unsatisfied or incomplete data,' 'unfavorable (meteorological) conditions' and so on will occur. Several regulations may be drafted and fed into the program through a different exception management process to manage these situations properly. Piggybacking in the error message corrective behavior adds value to the user experience. Hiding fine information, your user error message should provide an objective summary of the problem and a recommendation for the next best action(s).

This provides another machine learning opportunity whereby a trained model can predict behavior based on the variables and contextual data to follow.

3 Robotic Process Automation in Industries

It has always been a challenge to run a company that has high productivity and low cost. Complex procedures, constantly evolving regulatory standards, labour-intensive operations, unmanageable amounts, faster marketing time-these are some of the main challenges most financial institutions face. Operational complexities have reached a condition that they cannot be overlooked any longer. Automation is the secret to success in this dynamic world-automation that is easier, more effective and more scalable. Robotic Process Automation (RPA) suits the process automation requirement perfectly, as it offers low-cost solutions, minimal disruption, short payback time and high performance. Companies across the globe know that RPA is the next big digital transition that will enable workers to avoid working on repetitive tasks. RPA lets workers concentrate on more value-added projects that are crucial for the company's bottom line (Fig. 3).

Technological innovation not only impacts the company's organizational science but also impacts the human resource environment as the functions of human workers

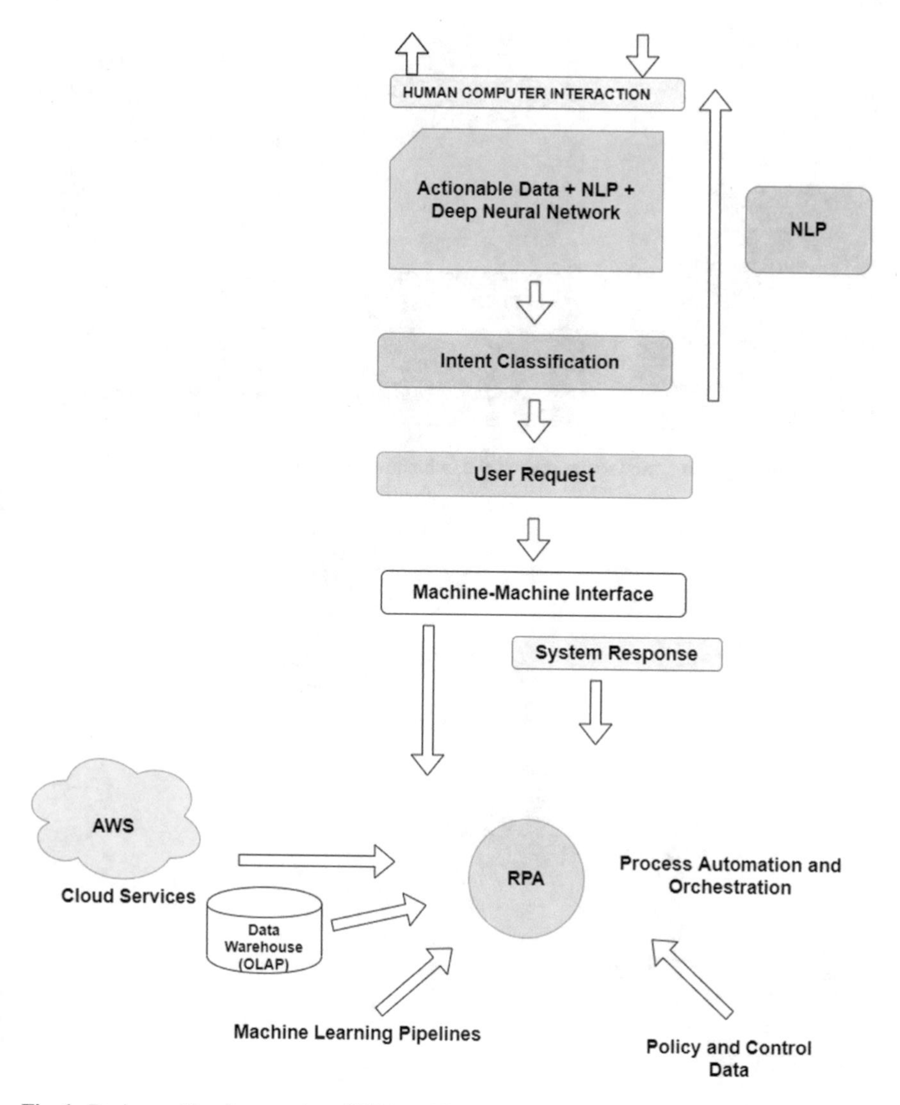

Fig. 1 Design and implementation of RPA + AI

are modified to suit the new technological paradigms. And while the technological aspects of RPA and AI are especially fascinating, one should not neglect the possible consequences that its development might have for the workplace and how it can rewrite other employees' positions. Parallels can be drawn to the effect of industrial automation on factory workers, as a consequence of which jobs in low-cost countries declined [4]. The effect on offshore outsourcing is one widely ignored aspect of increased automation. It is reported in an industry report [5] that in certain situations, RPA robots can now cost four times more than outsourcing the job to a low-cost

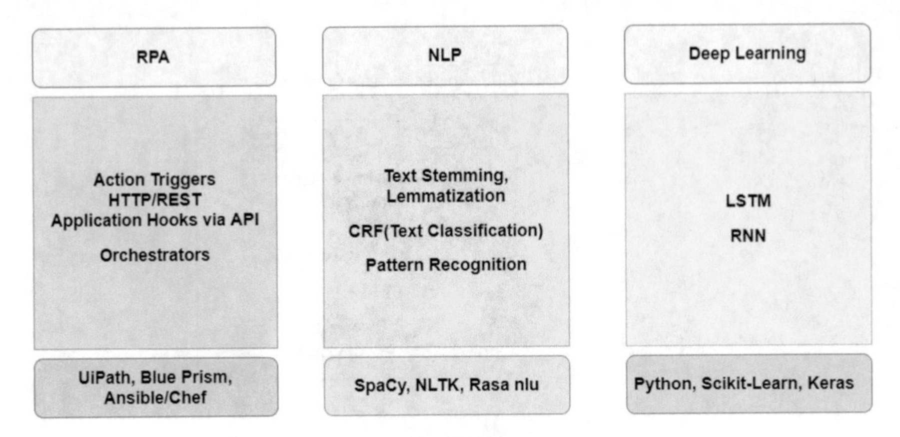

Fig. 2 Technologies required for a successful combination of RPA + AI (cf. [23])

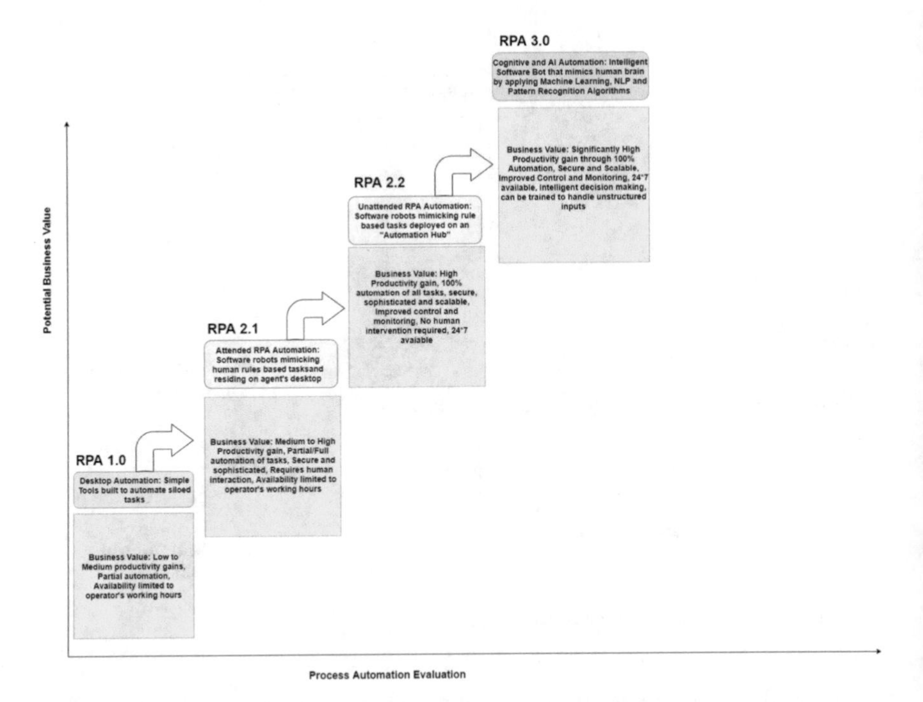

Fig. 3 Timeline of growth of RPA (cf. [24])

country. Though it is possible to dispute the exact cost ratio between robot and outsourcing, the essence is that increased automation will bring disruption to the global trend of overseas outsourcing (Fig. 4).

This insight is also supported by Wright and Schultz [6], who says that cheap labor providing countries like India and China will experience heightened hardship

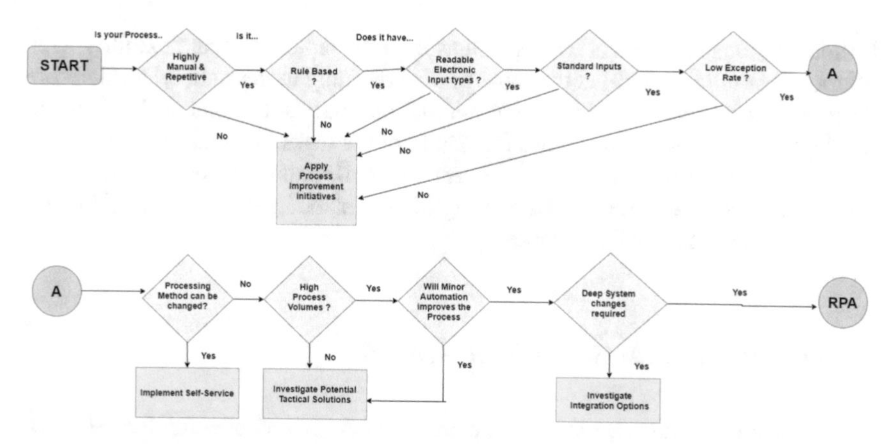

Fig. 4 An illustration guide of the selection of processes and working for Robotic Process Automation by using the principles of Robotics

as the need for cheap labour decreases as efficient automation becomes available. According to Manyika et al. [7], as a result of the developments in technology, some economists are worried about a "premature industrialization" in developed countries. On the other hand, automation may be necessary for more developed countries such as Japan to offset the declining labor supply resulting from their older population. For countries with a declining workforce, future economic growth depends on higher productivity and quality, whereas developing countries like India expect higher future GDP growth from an increasing labour force. Automation improves productivity and mitigates the problems of an ageing population while in developed countries it threatens job creation. This further strengthens the argument that automation benefits developed countries and limits developing countries' economic growth based on outsourcing tasks [6].

3.1 Polarization of Labour Market

Automation is argued to contribute to employment fragmentation as the provision of middle-skilled workers such as clerical work is decreasing while low-skilled and high-skilled jobs are rising [8]. The labour market division may lead to a mass migration of medium-skilled workers to more low-skilled jobs, while high-skilled workers with little to no routine work would benefit from automation. They found increased work polarization in all developing countries surveyed in a survey by the authors [8]. It is thought that polarization further depresses the earnings of people with little to no employment, thereby increasing the income gap in developed countries. Other researchers argue that right automation could very well turn manual routine jobs into highly skilled positions allowing employees to stay on as robot co-developers by, for example, maintaining and overseeing the AI algorithms through pseudo-code. It is thought to inspire the workers as well as improve organizational flexibility while offering cost-efficiency [9] at the same time. According to these theories, therefore, the growing need and use of intelligent automation would not automatically eliminate employment, but instead may shift the nature of the tasks and make middle-skilled employees migrate to high-skilled jobs rather than the contrary. Regardless of direction, both perspectives reinforce the theory of automated intelligence which induces the polarization of the labour market.

4 Human Intervention Is Still Needed

To offer some accurate overview of the effects of AI and automation for the workplace, it is important to decide what really is feasible or not. We are not yet close today to Artificial General Intelligence (AGI*)*, which is the terminology for the scenario where machines can do as well as a person any intellectual activity. In comparison to the satisfying AGI, effectively applied AI provides more solutions for a collection of business process tasks and problems, which is often referred to as "narrow" AI. And for now, there is no need to worry about machines removing the need for human thought. Despite this, AI technology currently available is still a base for digital disruption. Technological developments in automation and AI can constantly change labour markets that impact their major stakeholders such as employees, companies, government and society. In reality, they will rely on automation to a greater extent for companies to maintain their competitiveness in order to increase performance. Nevertheless, there is a risk of overlooking long-term macro effects of automation, such as fading mutual goodwill between business and stakeholders in favor of short-term financial winnings, with the need to remain competitive in performance. It is important for businesses to strike a balance between automation and human capital in order to be competitive in the long run.

5 Basics of RPA

Robotics Process Automation (RPA) allows businesses to automate processes across applications and systems, just as a person was doing them. RPA's aim is to move execution of the process from humans to bots. Robot automation integrates with the current IT infrastructure and needs no complex system integration. What separates RPA from other process automation is the unchanging existence of the underlying information systems [10]. Consequently, if a person should be reestablished to perform the task, there will be no problems because the user interface and the information systems are intact [11]. Initially, the RPA code was based on basic [if, else, other] statements, while the numbers of firms and researchers recognize emerging market opportunities by constantly expanding the available code library to encourage more sophisticated task automation. RPA is ideal for repeatable, well-documented, and very well-defined tasks. When the task is dependent on guidelines, and it doesn't change regularly, then it is a task that RPA should complete. RPA can automate a wide variety of tasks in many different industries. Here are just some of RPA's useful applications.

Scraping websites: RPA can be used to collect data from web pages. Examples of this include downloading and summing data from databases for stock trading. If the data is collected and analyzed, it can then be passed on for further study to humans. Automated email scanning: Lots of emails are received by many organizations asking the same questions. RPA can handle some of those emails and respond with common responses. The emails which can not be addressed by the RPA bot can then be forwarded to the correct responding workers. Cleaning data is a perfect example of where RPA can be used to achieve time-consuming assignments. RPA will sort out bad data even more effectively than humans if there are specific rules as to what constitutes bad data.

Data Entry: One of RPA's most comprehensive functions is that of data entry. An RPA bot can use optical character recognition (OCR) to read original types, and then "key" the data to an application. It would be quicker than a human being could and would be more precise in keying the data.

5.1 TA versus RPA versus Cognitive RPA

Test Automation—Automating the testing process is the procedure, i.e. creating a code or program to perform the manual testing task. There are various types of software on the market for this, both open source and payable. The main goal is to create a high-quality product by investing less time in testing.

Robotic Process Automation—The method of automating business processes in sectors such as telecom, healthcare, banking and so on. Using the robots program to carry out the tasks. There are various types of RPA tools in the market for this, such as UiPath, Automation Anywhere, Blue prism etc. The main goal is to simplify the business process with a commitment to reducing physical labour costs and time, without losing precision. Cognitive RPA requires human intervention/RPA processing orders. Using Google Assistant for example to find the shortest path while driving is a case of Cognitive RPA. It is important for both RPA and test automation to allow for quick development and updating of robust UI automation and API automation. Even, there are some major variations. RPA requires orchestration of an enterprise-grade type, high availability and emphasis on development. Test automation includes skills such as designing test cases, virtualizing the service, managing the test data, etc. Test automation is more complex in many ways than RPA. RPA allows you to decide how automation can be used to advance established business activities. With test automation, you need to decide which combination of factors could break an end-to-end process, and you also need to get, configure, and manage the test data and test environments required to make sure your tests are accurate and practical.

RPA tools like UIPath can be used to test the automation with Blue Prism. Vice versa, advanced automation applications, such as UFT, can be used with RPA if you have good coding skills. But RPA tools are better suitable for work.

5.2 *How to Implement RPA*

5.2.1 Due Diligence

Selection of tools is a part of due diligence. It is the most important level since the RPA-related benefits are directly correlated with how fine automation is to be performed. Any tool which is a good match for the project will build good scripts. Organizations will investigate the viability of processes and financial modelling so they can determine the ROI and complete their range of resources based on the tests. Commercial RPA tools in the market today include, among others, Automation Everywhere and BluePrism. There are also free/community versions for companies with small budgets, or process automation needs, such as UiPath and WorkFusion. There may also be cases that don't need any of these resources at all, and some of the commonly used scripting languages, including Visual Basic macros or Python, may instead achieve process automation. These cases are, however, very rare, as these languages need additional scripting efforts compared to standard RPA devices, and thus may not be the preferred option for most organizations.

Organizations need a proof of concept (POC) to assess if the systems are technically feasible. When the POC has been established, companies will conduct a cost/benefit analysis to assess the potential benefits of implementing automation. This analysis should be explicitly quantified since the ROI break-even relies on a specific cost/benefit analysis.

5.2.2 Risk Identification

A further type of due diligence is process selection. The crucial juncture of process selection is the stage for evaluating candidates for automation. Those processes which are preferred candidates for automation meet the following requirements: (i) Stable Processes, (ii) Repetitive in Nature, and (iii) Organized in Nature. RPA's effectiveness depends on the selection of certain processes to be built and implemented. Organizations have to realize that it is difficult to optimize 100%. Therefore, which processes are chosen and which are rejected is a deciding factor for success with RPA.

5.2.3 Bot Creation and Dry Run

If a tool has been chosen by an organization and processes identified, your process steps should be identified which require a thorough review. The company will build

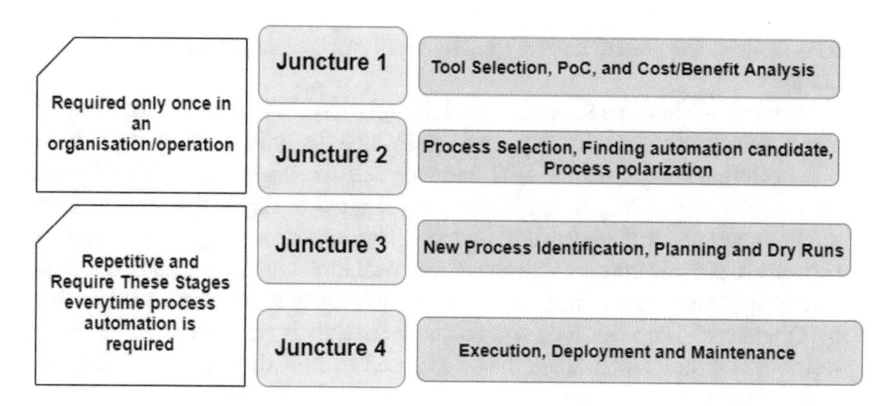

Fig. 5 Overview of the stages performed for Bot deployment

a detailed plan in this modeling which includes the exact costs. As in the preceding phases, threats are identified to minimize those in the construction effort and cost increases. This plan will explain exact breakeven durations and ROI values, including cost incurred as input while the automatic flow is produced, and cost saved during automated flow execution.

5.2.4 Execution and Maintenance

This is the final stage where the bot is deployed for execution. The changes need to be maintained if the processes are dynamic in nature. Figure 5 below shows a summary of all the processes.

6 Challenges of RPA

As we can see, RPA provides a variety of benefits that allow businesses like our hypothetical insurance company to automate much of their business operations and provide excellent, quick and reliable customer support.

All have two sides and thus RPA is not without obstacles. Without such problems, there is no way to incorporate new technology into the existing architecture of an organization. Here are only some of the obstacles that could stand in the way:

- Typically RPA is a short-term/temporary solution. Sure, it can offer instant relief and gain, but it requires complex omnichannel systems and regular analysis and optimization of the process in the long run.
- Current RPA tools available on the market have no or minimal capacity to learn about the system—we are still very far from the moment when automation is genuinely cognitive. With AI and computer technology being incorporated with

RPA devices, the potential for automation will increase exponentially, but we are not there yet.

- Ownership—Which means who owns RPA solutions. Is it IT or Business departments driving it? The company needs to include the criteria, support the feasibility solution design, help in UAT and then calculate the success rate. IT teams have a limited function which is primarily limited to supporting infrastructure requirements and testing data development. Business does not generally have the capacity to provide the comprehensive technical level requirement at BRD (Business Requirement Documentation) time or to identify test scenarios as required by QA teams. The need for good business analysts is paramount, but they are restricted on-the-market talent that is exposed to RPA design and solution and understands potential value.
- Limited Use—This can include handwritten papers. Partial automation based on an analysis of costs and benefits may be considered if full end-to-end automation is not feasible or reachable within a realistic timeframe.
- Change Management—Business and IT organizations will work together to proactively provide RPA support team with the program to company changes to upgrade scripts once they are in production. When several applications are used in the process it can pose additional challenges. Any improvement in the front end UI will hence have an effect on the RPA script, the output.
- Employee resistance and on boarding—This is one of the greatest issues, and it affects all businesses because of a basic fact—people are scared of nature shifting. Any changes that follow a new technology implementation may be disruptive to workers because they can encounter adjustments in their roles. For successful adoption, regular contact from business leaders and corporate sponsors to ensure that workers are fully aware of what is required of them in the implementation process is crucial. Promoting an innovation culture within the organization would only intensify this acceptance further, and remove behavioral resistance.

7 Artificial Intelligence

John McCarthy coined the term Artificial Intelligence at the Dartmouth Conference for the first time decades ago in the year 1956. He described AI as "the science and engineering of making smart machines." Artificial Intelligence, in other words, is the science of having machines to think and make decisions like humans. AI has been able to achieve this in the recent past by developing devices and robots that have been used in a wide range of fields including education, engineering, marketing, business analytics and many more. The major types of AI are discussed below.

Commonly known as weak AI, Artificial Narrow Intelligence means applying AI only to particular tasks. Currently, the latest AI-based systems claiming to use "artificial intelligence" work as a poor AI. Alexa is a perfect example of narrow AI. It operates within a restricted range of predefined functions. Alexa doesn't have any real intellect or self-confidence. Google's search engine, Sophia, self-driving cars and even the popular AlphaGo, fell into the poor AI group. Artificial General Intelligence Commonly known as Strong AI, Artificial General Intelligence comprises computers capable of performing any intellectual function a human being might perform. You see, computers don't have human-like skills, they have a solid processing unit that can perform high-level computations but they aren't yet able to think and reason like a human being. Many experts doubt that AGI would ever be feasible, and many even question whether it will be beneficial. For example, Stephen Hawking warned: "Powerful AI will take off alone, and re-design itself at an ever-increasing pace. Human beings, constrained by slow biological evolution, have not been able to succeed and will be superseded." Artificial Super Intelligence is a term that refers to the moment when machines can exceed human capabilities.

Today, ASI is seen as a hypothetical scenario as portrayed in films and science fiction novels, where robots take over the planet. But tech masterminds like Elon Musk claim that by 2040, ASI will take over the planet.

7.1 Major Concepts of Artificial Intelligence

Since AI is a broader field, below is the description of the major AI Concepts related to our work [5].

- **Machine Learning**—It is that part of AI that uses statistical techniques to help machines learn through experience.
- **Neural Networks**—It is the subset of ML which adds complexity to the statistical methods through the use of multi-layer networks. Neural Networks is inspired by the human Neural Networks in the brain and consists of several layers of nodes that together form a model that can be trained [12].
- **Natural Language Processing**—Umbrella term for the interaction between human language and computers. Historically, it has been done through rule-based programming, but today it is most commonly done through the use of ML methods, for example, Neural Networks [13].
- **Natural Language Understanding**—Not to be confused with its umbrella term NLP. NLU refers to the specific task of transforming unstructured data into structured data. NLU is the technique for a machine to interpret human language [13] and is, for example, used when Apple's Siri or Amazon's Alexa tries to understand user requests.
- **Natural Language Generation**—Not to be confused with its umbrella term NLP. NLG is the technique for machines to generate human language based on available data and can be seen as the reverse process of NLU. NLG is for example used when Siri or Alexa answers user requests.
- **Optical Character Recognition**—Umbrella term for techniques that enable an input image to be interpreted by a computer. Modern OCR techniques include ML algorithms, such as Neural Networks, in the recognition process [14].

- **Supervised Learning**—Supervised learning is the field of machine learning where input and output are given to the system with the goal of finding patterns for connecting these. Data classification is established in supervised learning, so the goal is to find a function that matches the correct classification [15].
- **Unsupervised Learning**—Unlike supervised learning; unsupervised learning is the learning where input data is given to the computer during preparation. In unsupervised learning, the classification of data is unknown and it is often unknown how many classes that should be considered [16]. One technique that makes use of unsupervised learning is clustering.
- **Clustering**—Clustering is a technique that looks for patterns and sequences in data [16]. These patterns are often undetectable by humans, commonly because of the sheer amount of data or because the patterns might not be logically intuitive. Clustering techniques could be used in a similar way to regression analysis, with the advantage that variables and their interactions do not have to be defined in advance since the computer itself can find relevant patterns in the variables. The outcome of a clustering algorithm is a grouping of subjects depending on their (almost) undetectable characteristics.
- **Classification**—A classification problem is aimed at categorizing an object based on certain attributes [17]. This can, for example, be done through a Neural Network, where the final layer represents the category for the object. Classification can also be done without AI algorithms, but would then only take naive considerations to defined parameters and not see patterns that are not detectable by humans.
- **Auto Speech Recognition**—Automatic speech recognition, or as more popularly referred to, Speech recognition, is the operation of transforming spoken words into written text [18]. Note that speech recognition does not interpret the meaning of the content, just as OCR does not analyze its input. To interpret speech, NLP can be utilized after the speech recognition process is done.

7.2 Limitations of Artificial Intelligence

Knowledge of AI's possibilities opens up great opportunities but understanding AI's weaknesses are equally important. The authors in [19] note that there have traditionally been times in which the progress of AI has frozen, called AI winters. Similar to the impact of the automation hype described in the RPA segment, as standards are set too high, AI hype often has negative consequences that the technology simply can not live up to. They also note that AI excitement followed all of the preceding AI winters. Better education can be fairly manageable by inadequate information about AI limitations, Obstacles which could be more difficult to climb, including regulatory barriers. We claim that regulatory concerns are one of the biggest potential impediments to AI growth. This challenge becomes especially important for the financial services industry applications which involve a lot of personal data. The problem of training a Neural Network is one specific example that is implemented.

Can customer data be used to train the Neural Network or do we have to use produced data, and if we use produced data that is not "true," how does the Neural Network manage the discrepancy? Burgess [20] recognizes that a growing issue in this field is that regulators continue to fall behind and there is a possibility that regulations will be implemented which will cause certain applications to fall into the legal landscape's grey zone. He argues that AI should be more embraced in time and then enforced by regulators, but in the meantime, businesses need to be ready to provide answers and clarity about their AI and data practices. Authors in [19] also explore the possibility of unexpected regulatory changes that could shock the industry and disrupt it. Taddeo and Floridi [21] note that *AI4People* was created by the European Parliament, which is an e-purpose to direct AI towards people's good. The European Commission also has an AI network of experts who can support lawmakers on related matters. Taddeo and Floridi [21] argue that moral responsibility needs to be shared out among designers, regulators and users. Most AI techniques demand a lot of data. Plastino and Purdy [22] claim that AI's efficiency depends directly on the consistency and the amount of available data. The algorithms may be in place in some situations, the company may be ready, but the data does not exist, is not available or is rather unstructured.

8 AI-Based Cloud

In the present era, cloud computing has emerged as a common aspect of modern software systems. The cloud computing industry has now become a battle that is dominated by the four tech giants overwhelmingly: Amazon, Microsoft, Google and IBM with a handful of other cloud services in different markets, such as AliCloud in China. It's hard to conceive about a company being innovative enough to challenge the current dynamics of such a concentrated industry. Artificial Intelligence is the kind of technology that is capable of not only developing current established cloud systems but also driving a new generation of cloud computing innovation. It is an intriguing idea to say that AI would affect a new generation of cloud computing infrastructure, given that convergence technology advances such as mobile or the Internet of Things (IoT) have not transformed the cloud computing environment. However, the analysis makes sense if we add movements such as the Smartphone or IoT and AI to a substantial difference. In this context, cloud providers were not meant to provide the runtime for running IoT or mobile applications, but rather services which would require backend functionality for those solutions. For e.g., the next-generation cloud AI platform, supplied on request with optimum GPU capabilities, should be able to deploy an authored system using a deep learning framework such as Tensor Flow or Torch across hundreds of nodes. AI-first cloud is a paradigm of cloud computing of the next decade, designed around the capabilities of AI. While we do not know exactly how the design of AI-first cloud platform should look, we still have some interesting thoughts to explore:

- Optimized GPU facilities: Modern cloud facilities would need to support GPU environments designed for fast computing in order to run arbitrary complex AI processes. With Microsoft announcing the launch of the N-Series GPU instances as part of the Azure platform, we are already seeing some early efforts in this regard.
- AI-first network infrastructure: The next step of cloud computing solutions will go beyond supporting AI applications innovations and using AI in its application and web infrastructure as a first-class citizen. We can see AI in the future as a key element for enhancing intelligence in cloud services like space, computation, or security.
- Combination with prominent PaaS services: The next generation of cloud computing platforms will ensure seamless integration between AI and deep learning systems and the current catalog of cloud services used in cloud platforms, in order to create advanced AI applications.
- Support for popular AI frameworks: Cloud computing systems of the next generation should be able to run deep learning or AI applications implemented in conventional frameworks such as Tensor Flow, Caffe, Theano, Torch, etc. in the same way they now support the implementation of web applications or background methods. At that point of view, the AI-first cloud should not be limited to a single AI platform but should support heterogeneous, deep learning systems that developers around the world are currently using.
- Managing tools: Capacity constraints in operational technology are one of the main challenges facing the new generation of deep learning and AI. Cloud computing systems of the next generation are in a unique position to address this challenge by offering advanced tools to track and run the AI programs that are deployed within their network.

Cloud computing is a very well-known trend of technology that is largely dominated by companies such as Amazon, Microsoft and Google. From that point, it would seem impossible to predict a technological change that would threaten today's cloud computing landscape. But AI brings with it some very unique features which will definitely affect the next generation of platforms for cloud computing. AI needs a modern computing infrastructure which promotes completely new programming paradigms and frameworks. We would expect the cloud's incumbents to incorporate AI capabilities in the near future as a core element of their infrastructure and then we'll see a new wave of AI-driven cloud emerging.

9 Robotics 2.0

With the rise of the Internet, autonomous robots came to be already in the '90s. Scrapping and spamming robots are popular on the web today. In Robotic Process Automation systems, more industry-oriented robots are designed to perform routine tasks in offices—from copying data to spreadsheets, coding, and monitoring emails.

You just show them how you do it step by step and they do it over and over again, mindlessly. The design of these machines takes time, but if you have a repetitive, tedious job performed every day by hundreds of people, the time is most often worth it. Just like it is worth having humans replaced by automated weapons in factories, so the entire manufacturing room is safer as robots are given routine, physical tasks.

This first generation of autonomous robots is an automated version of mechanical equipment in factories—they have no 'vision' or 'sensing' and can only perform the same type of action they have been trained to do without changes to changing situations or any sign of creativity. In the rise of AI, autonomous robots end up 'seeing' incoming data and 'saying' about it. They don't have to blindly follow the rules but they can improvise on the spot. Digital Robots will eventually become Conscious. Technologically speaking, machine learning—especially in-depth learning and learning enhancement allows for great stability in RPA systems. However close it can sound to Artificial General Intelligence, which is still far away, these kinds of smart digital robots will soon flourish in all business verticals (and typically intelligence verticals), where one works on extremely restricted dictionaries/ontology's e.g., tax definitions, financial reports, or legal agreements. Right now, we're at the very beginning of this new movement, reducing routine, dull tasks and enabling people to focus on truly creative work. The transition would occur in professions that seemed out of reach of automation at first: consultants, bankers and lawyers. Most workers will change significantly over the next 5–10 years.

9.1 Data Management with Respect to Robotics 2.0

One of the greatest drawbacks of AI implementations, as previously stated in the AI section, is access to data that is appropriate for the task in hand [22]. It is important, in order to navigate the automation process that organizations have an organized way of managing their data and everything that comes with it. A pervasive 2018 states that data and exposure to it is one of the three most serious obstacles to Robotics 2.0 implementation. Many large organizations are adding a Chief Data Officer to their C-Suite management, and according to Gartner (2016), this will be in place by 2019 at 90% of large companies. According to Plastino and Purdy [22], however, the Chief Data Officers primarily deal with laws, data security, and governance. They argue that businesses need to appoint a Chief Data Supply Chain Officer, who can create a seamless end-to-end data supply chain, in order to fully leverage the benefits of data in AI.

9.2 Applications of Robotics 2.0

There are many applications based on Robotics 2.0 but we'll discuss some major sectors that will get totally advanced with the help of this technology.

- Healthcare—An efficient RPA will lead to better care delivery, with major changes in the quality of coordination of treatment, population health, remote monitoring and control of use. If healthcare workers are no longer overwhelmed with many of their everyday duties, they have the ability to concentrate on more important things such as one-on-one contact with patients. Similarly, cases in which AI-assisted image recognition is implemented in order to identify malignant traces can be combined with a warning and notification system so that sensitive issues such as cancers can be tumored before calling out by the domain experts.
- Accounting—There are various end-to-end processes within accounting, such as Procure to Pay, Order to Cash and Record to Report cycles, which can be heavily automated via RPA. Many businesses have simplified their invoicing process for a particular example, but on-boarding of new suppliers is often performed manually. Implementing RPA will screen new suppliers and produce a complete report on their credit ratings, tax details, etc.
- Human Resources—RPA may boost activities such as accounting, registration of benefits, on boarding, and recording of compliance that all entail a large amount of manual, repetitive labor. At Ernst and Young, RPA is used to monitor the continued education of staff, and inform the HR department if further training needs further intervention.
- Automotive Industry—Previously time-consuming, laborious, and menial activities, such as emailing customers and products, digitizing documents, handling orders and payment transactions, as well as controlling procurement processes, could now be performed by RPA software robots with the click of a button. Automation technologies also allowed real time supply and demand monitoring. This meant that consumer demand with production capacity and inventory rates could be more effectively matched, and that goods could be delivered to consumers at competitive prices. In introducing RPA, the car manufacturer concerned has been able to maximize market efficiency in achieving learner operations, better inventory management and resource procurement, and enhanced coordination with suppliers and clients. What's more, RPA allowed the manufacturer to more efficiently assign its employees to jobs which allowed them to serve customer interests better.

10 Conclusion

The pace at which the RPA and AI are growing, the market will definitely look brighter in the years ahead. Also giant firms such as IBM, Microsoft and SAP are gradually tapping into RPA. In other words, they increase the visibility of RPA software and its momentum. Moreover, new vendors are also emerging at a rapid pace and have begun to mark their presence in the industry. However, the talk is not just about RPA, the role of AI is one of the most important things at the moment as well. Robotics 2.0 concept is gaining popularity among many organizations. The industry is now witnessing their capabilities and why not AI can read, listen, and analyze, and then feed data into bots that can produce, bundle, and send output. RPA and AI are, at the end of the day, two useful techs that companies can use to help the digital transformation of their company. Looking to the future, it seems that the talks of computers being the sidekick of humans would be more than talks of RPA and AI taking human jobs off. To boost product efficiency and thoroughly develop practical accuracy, clear understanding of conversational AI is a must. Machine learning pipelines are the fundamental building block for delivering reliable outcomes. Many find RPA as a much needed aspect of digital transformation. If software robots are used along with AI and driven by policies and historical data, it is possible to achieve productivity improvements, reap the benefits of automation, AI/ML and build opportunities for idyllic user experience, raise user engagement levels, gain operational excellence and increase ROI.

References

1. https://www.gartner.com/en/newsroom/press-releases/2019-06-24-gartner-says-worldwide-robotic-process-automation-sof
2. Mohanty S, Vyas S (2018) It operations and ai. In: How to compete in the age of artificial intelligence. Apress, Berkeley, CA, pp 173–187
3. Chui M (2017) Artificial intelligence the next digital frontier?, vol 47. McKinsey and Company Global Institute, pp 3–6
4. Targowski A, Modrák V (2011) Is advanced automation consistent with sustainable economic growth in developed world?. In: International conference on ENTERprise information systems. Springer, Berlin, Heidelberg, pp 63–72
5. Bellman M, Göransson G (2019) Intelligent process automation: building the bridge between Robotic Process Automation and artificial intelligence
6. Wright SA, Schultz AE (2018) The rising tide of artificial intelligence and business automation: developing an ethical framework. Bus Horiz 61(6):823–832
7. Santana M, Cobo-Martín MJ (2020) What is the future of work? A science mapping analysis. Eur Manag J
8. Goos M, Manning A, Salomons A (2014) Explaining job polarization: routine-biased technological change and offshoring. Am Econ Rev 104(8):2509–2526
9. Kopeć W, Skibiński M, Biele C, Skorupska K, Tkaczyk D, Jaskulska A, Abramczuk K, Gago P, Marasek K (2018) Hybrid approach to automation, RPA and machine learning: a method for the human-centered design of software robots. arXiv preprint arXiv:1811.02213
10. Van der Aalst WM, Bichler M, Heinzl A (2018) Robotic Process Automation
11. Asatiani A, Penttinen E (2016) Turning Robotic Process Automation into commercial success—case OpusCapita. J Inform Technol Teach Cases 6(2):67–74
12. Shanmuganathan S (2016) Artificial neural network modelling: an introduction. In: Artificial neural network modelling. Springer, Cham, pp 1–14
13. Deng L, Liu Y (eds) (2018) Deep learning in natural language processing. Springer
14. Phangtriastu MR, Harefa J, Tanoto DF (2017) Comparison between neural network and support vector machine in optical character recognition. Procedia Comput Sci 116:351–357
15. Hudson DL, Cohen ME (2000) Neural networks and artificial intelligence for biomedical engineering. Inst Electr Electron Eng
16. Kasabov NK (2019) Time-space, spiking neural networks and brain-inspired artificial intelligence. Springer, Heidelberg
17. Amezcua J, Melin P, Castillo O (2018) New classification method based on modular neural networks with the LVQ algorithm and type-2 fuzzy logic. Springer

18. Renals S, Hain T (2010) 12 speech recognition. In: The handbook of computational linguistics and natural language processing, vol 57
19. Lacity M, Willcocks LP (2018) Robotic process and cognitive automation: the next phase. SB Publishing
20. Burgess A (2017) The Executive Guide to Artificial Intelligence: how to identify and implement applications for AI in your organization. Springer
21. Taddeo M, Floridi L (2018) How AI can be a force for good. Science 361(6404):751–752
22. Plastino E, Purdy M (2018) Game changing value from artificial intelligence: eight strategies. Strategy Leadersh
23. Morathi LP (2020) Millennial perceptions of the 4th industrial revolution in an information technology company. Doctoral dissertation, North-West University, South Africa
24. https://www.capgemini.com/2019/11/robotic-process-automation-an-industry-perspective/#

Printed in the United States
by Baker & Taylor Publisher Services